SUPERNOVAE

ASTROPHYSICS AND SPACE SCIENCE LIBRARY

A SERIES OF BOOKS ON THE RECENT DEVELOPMENTS
OF SPACE SCIENCE AND OF GENERAL GEOPHYSICS AND ASTROPHYSICS
PUBLISHED IN CONNECTION WITH THE JOURNAL
SPACE SCIENCE REVIEWS

VOLUME 66
PROCEEDINGS

SUPERNOVAE

THE PROCEEDINGS OF A SPECIAL IAU SESSION ON SUPERNOVAE
HELD ON SEPTEMBER 1, 1976 IN GRENOBLE, FRANCE

Edited by

DAVID N. SCHRAMM
Department of Astronomy and Astrophysics
and the Enrico Fermi Institute,
The University of Chicago

Jointly sponsored by

IAU COMMISSION 35 ON STELLAR EVOLUTION AND
COMMISSION 48 ON HIGH ENERGY ASTROPHYSICS

D. REIDEL PUBLISHING COMPANY
DORDRECHT-HOLLAND / BOSTON-U.S.A.

Library of Congress Cataloging in Publication Data
Main entry under title:

Supernovae.

(Astrophysics and space science library; v. 66)
Bibliography: p.
Includes index.
1. Supernovae–Congresses. I. Schramm, David N.
II. International Astronomical Union. III. Series.
QB835.S94 523.8'446 77-1178
ISBN-13: 978-94-010-1231-7 e-ISBN: 978-94-010-1229-4
DOI: 10.1007/978-94-010-1229-4

Published by D. Reidel Publishing Company,
P.O. Box 17, Dordrecht, Holland

Sold and distributed in the U.S.A., Canada and Mexico
by D. Reidel Publishing Company, Inc.
Lincoln Building, 160 Old Derby Street, Hingham,
Mass. 02043, U.S.A.

TABLE OF CONTENTS

PREFACE

Supernovae are among the most exciting things occurring in the universe. Much recent research has concentrated on phenomena related to supernovae. For example, the origin of the cosmic rays and the origin of the bulk of the heavy elements seem to be closely associated with the phenomenon of supernovae. With the discovery of the pulsar in the Crab, it seemed clear that supernovae were also intimately associated with the formation of neutron stars and perhaps even black holes. The purpose of the conference, of which this volume contains the proceedings, was to bring together the leaders of supernova research, each of whom has concentrated on different aspects of the problem, to try to form a coherent picture both observationally and theoretically of our current understanding of supernovae. In so doing, key invited talks were presented on the light curves of supernovae, both observationally and theoretically; on the possible uses of supernovae, for example in determination of the Hubble Constant; on the formation and evolution of supernova remnants, again both observationally and theoretically. The possibility that supernovae might explain quasars was also presented. A review of the current status of statistics of supernovae was presented, giving the rate at which they go off and the implications with regard to what mass stars are the progenitors for supernovae. Again, this was presented both from the observational point of view and from the theoretical stellar evolution point of view. There was even mention of how supernova explosions might stimulate star formation and perhaps even the formation of the solar system itself. The nucleosynthesis that occurs in supernovae was discussed and the session ended with detailed discussions of mechanisms which may provide an explanation of how supernovae actually do explode. The conference was held in conjunction with the International Astronomical Union meeting in Grenoble, France. The date of the special session was September 1, 1976.

It is believed that this session was probably the most up-to-date comprehensive session on supernovae presented anywhere to this time. The session is unique in that it did not concentrate just on the observational aspects nor just on the theoretical aspects. Similarly, it did not concentrate just on the interior mechanisms of the explosion nor just on the exterior phenomena. It did not concentrate merely on the remnant nor on the hot dense cores. As a result, this volume and the conference it represents form a very concise summary of the complete phenomenon of supernovae.

The list of authors for the volume includes most of the leading researchers in supernovae research. In particular, Leonida Rosino discussed the supernova observations and presented the work of the Asiago Observatory, which has been one of the leaders in the field of gathering information on supernova light curves. Gordon Lasher in collaboration with Karp and Chan presented his work on theoretical supernova light curves. In particular, it was shown how a type I supernova light curve may depend on the hydrogen in the envelope of the presupernova star. David Branch, who has worked extensively on supernovae while he was at Hale Observatories and at Herstmonceux as well as at his present position at the University of Oklahoma, presented his research showing how supernovae may be used to determine the Hubble Constant, in particular showing the strength of a supernova-based distance scale. This distance scale is no longer hypothetical but is now getting to the point where it may actually determine a precise value of the Hubble Constant. Leonard Culhane of the University College in London presented his group's very interesting work on x-rays from supernova remnants. This work has led to a great deal of understanding on how remnants evolve and on how the energy is distributed in the remnants. Roger Chevalier of the Kitt Peak National Observatory in his paper presented a summary of his work on the evolution of supernova remnants and in particular discussed the Crab nebula and its evolution and how it may have been a type II supernova. In addition to the theoretical work by Chevalier on supernova remnants, three observational papers on the evolution of supernova remnants were presented, one by Dickel, Wells, Gull, Willis and Van den Bergh, another by Weiler and Wilson, and the third by Gull, Parker and Kirshner. Each of these showed the development of new techniques for observing supernovae and how these new techniques are leading to a greater understanding of the remnants themselves. Stirling Colgate of Ward, Colorado and Al Petschek of New Mexico Institute of Mining and Technology presented their theory on how supernovae might explain the quasar phenomenon. Colgate is one of the well-known leading theorists on supernova phenomenon, both on the interior supernova mechanisms and on the treatment of shocks emerging from the supernovae. In this particular paper, it is shown how these shocks and their interaction on the surrounding material might be able to be interpreted as providing a way to explain important features of quasar observations.

One of the highlights of the special session was a report by Gustav Tammann of Basel and of the Hale Observatories. Tammann presented a summary of his comprehensive work on the analysis of supernova statistics. In particular, Tammann examined what is known about the rate of supernova occurrence in various types of galaxies, including our own. His findings were in some cases contrary to many preconceived notions but were in all cases supported by careful observations. Beatrice Tinsley of Yale University presented a careful review of what is known about the masses of the stars which eventually become supernovae, in particular discussing what is known about the possible difference in the masses of stars which eventually become type I supernovae and those which eventually become type II supernovae. Again, what was said did not always agree with the previously held opinions but was certainly supported by sound scientific logic. Zalman Barkat of the Hebrew University in Jerusalem presented his work on the evolution of stars which eventually become supernovae. This is of course directly related to the work on supernova masses presented by Tinsley and of course this entire problem of which mass stars eventually become supernovae ties back into Tammann's work on the rate at which supernovae go off in various galaxies. Barkat's work on the presupernova evolution of stars showed how various uncertain quantities in the evolution of stars can drastically affect the configuration that they will have at the time that they become supernovae. In particular, the theory of convection and how it is applied in stellar evolution is extremely critical to this problem.

W. Herbst of the Carnegie Institute presented an observational case for supernova-induced star formation. This is particularly interesting in view of the recent work by Typhoon Lee, D. Pappanasstassiou and G. J. Wasserburg which showed that the radioactive nucleus ^{26}Al with a half-life of less than a million years was present when objects solidified in the early solar system. In order for this nucleus to have been present at that time means that a nucleosynthetic event must have gone off within a few million years of the solidification of objects in the solar system. The two possibilities for this nucleosynthetic event are (1) a supernova, and (2) an early irradiation in the early solar system. This latter possibility is plagued with many constraints, in particular the energetics required and the absence to date of other anomalies being observed, which would be expected from such an irradiation. This would tend to imply that at least one supernova did indeed go off within a few million years of formation of solid objects in the solar system, which would say that the formation of our solar system was closely related to a supernova and thus perhaps that supernovae did indeed stimulate the formation of our solar system. The nucleosynthesis that goes on in supernova explosions was discussed by James Truran of the University of Illinois. Truran, who is one of the leading experts on nucleosynthesis in stars, gave an exceedingly comprehensive review of nuclear astrophysics in exploding stars.

The mechanisms by which the supernovae actually eject their outer material were discussed in two presentations. The first, a summary of the Russian work, is included in this volume in a paper by Chechetkin, Imshennik, Ivanova and Nadyozhin. Due to a last minute change in traveling plans by Nadyozhin, this paper had to be presented at Grenoble by Stirling Colgate, who gave an excellent review of the paper on exceedingly short notice. The original paper as included in this volume presents work on carbon detonation supernovae and neutrino transport supernovae that is currently going on in the Soviet Union. The other paper summarizing the mechanisms for supernova explosions was presented by Richard Epstein of Harvard and NORDITA. Epstein has been a close collaborator of Schramm and Arnett and Cameron as well as having frequent conversations with James Wilson of Livermore. Therefore, Epstein was in a position to be able to present a complete review of work going on in the U.S. on mechanisms for supernova explosions. Epstein concentrated on the gravitational collapse of massive stars and discussed the most current problems in neutrino transport mechanisms as well as describing some of the problems associated with the equation of state and the hydrodynamic bounce.

From the above, it seems clear that this volume does provide a complete, up-to-date summary of supernova research and would therefore be a good starting place for anyone interested in finding out the current status of the subject.

I would like to thank Martin Rees for his work in helping to arrange for the session in Grenoble. I would also like to thank Jean Audouze for his help in coordinating the supernovae session with the CNO isotope session. In addition, special thanks is due to all of the authors for providing such outstanding manuscripts for this volume. Finally, I would like to thank Markie McRae, my secretary, without whose help this volume could not have been completed.

David N. Schramm
Editor

OBSERVATIONS OF SUPERNOVAE AT THE ASTROPHYSICAL OBSERVATORY OF ASIAGO

L.ROSINO
Institute of Astronomy of the University,Padova

Since 1959 a program of systematic search of supernovae was carried out at Asiago with the 40 cm Schmidt telescope.The survey, which at first was limited to 36 high galactic latitude fields, was later extended to 74 fields and brought to the discovery of 21 supernovae. Later on, in 1964 , when a second Schmidt telescope of 67 cm became available at Asiago, a new program concerning the determination of B and V light curves was initiated. At the same time spectra of the brightest supernovae were successfully obtained at the newtonian focus of the 122 cm telescope , with dispersions from 130 to 280 A/mm at H_γ . The use of image-tubes, after 1968, allowed us to extend the spectroscopic observations to supernovae as faint as magn.16. A complete bibliography of the supernovae observed at Asiago from 1957 to 1972, together with the principal results of the survey, is reported in the Proceedings of the Lecce International Conference (Rosino and di Tullio,1974).

Both programs, the search of supernovae and the study of their light curves and spectra, are still continuing. In the search program, carried out mainly with the 40 cm Schmidt, but occasionally also with the 67 cm telescope, all of the galaxies with more than one supernova have been included, with the purpose of obtaining new clues on the

David N. Schramm (ed.), Supernovae. 1-11.

problem of the supernova frequency. A third supernova was found in the galaxy NGC 4303 (Rosino, 1964) and a fifth in NGC 6946 (Rosino,1971).

The possibility of using a fairly large and homogeneous sample of SN-I light curves, all obtained with the same instruments under similar conditions, induced us (Barbon et al.,1973a) to verify whether it could be possible to derive for supernovae I an average light curve. The tentative was successful. The standard curve (Fig.1) obtained by the fitting of 38 type I SN, and the mean color curve

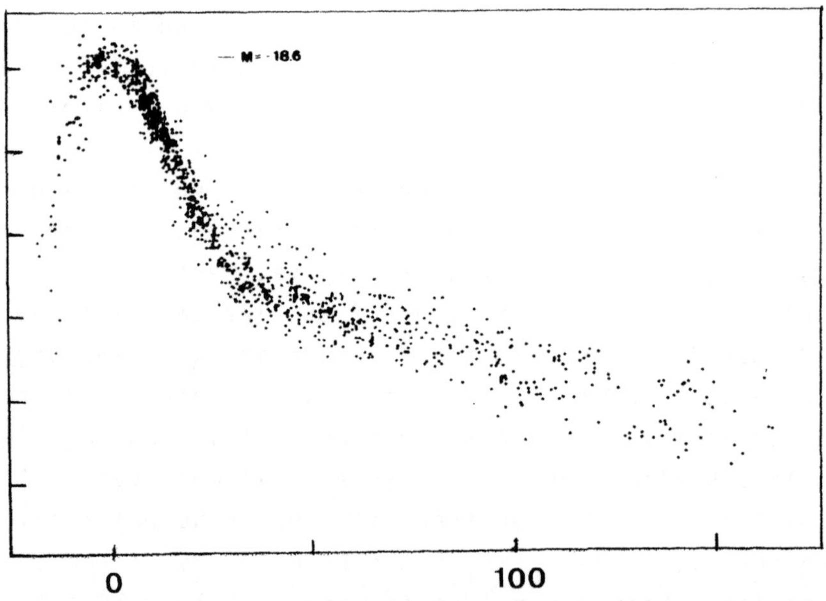

Fig. 1 -Average light curve of type I supernovae

have been of the greatest use for further studies on supernovae, at Asiago and elsewhere; they have offered the possibility of obtaining epochs and magnitudes of maximum and color excesses, even with a limited number of scattered observations.

Type I supernovae are also good distance indicators. So, it has been possible, in some cases, to derive the distance of the parent galaxy from the supernova brightness. At large distances, the redshift of the strong absorption feature at λ 6160 gives a still more effective method,independent on absorption, for obtaining the distance.

Particular attention was given to type II supernovae. These stars have generally a lower luminosity than type I supernovae (Kowal, 1968) and therefore a lesser probability of being discovered.Fairly good light curves are only available for 22 of them. Although these curves show some remarkable differences, yet it has been possible to derive from 17 of them an average light curve,characterized by an early decline, a plateau of nearly constant brightness

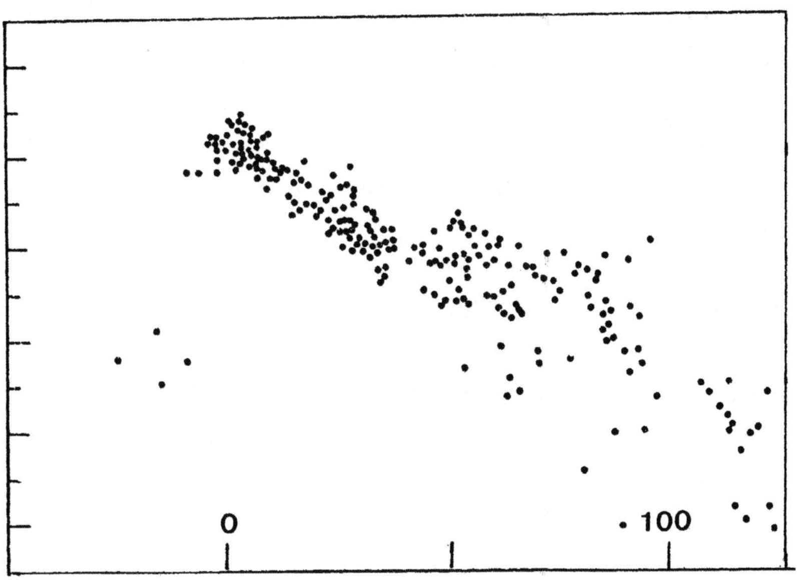

Fig.2 - Average light curve of 17 type II supernovae

and a final decline (Fig.2).The remaining five type II supernovae have light curves which do not follow the regular pattern and form two other groups with similar properties. In conclusion, the data are still insufficient and further observations of SN II, extended for more than 100-150 days, are requested.

We pass now to the supernovae recently observed at Asiago. Light curves and spectra of nine supernovae appeared between 1970 and 1972 have been already published by Barbon et al. (1973 b , 1973 c). The present communication gives some preliminary results obtained by Dr.F.Ciatti and the writer on other five supernovae, observed between 1972 and 1975.

SN 1972-j in NGC 7634 - This supernova was discovered at Asiago as a star of magn.15 on August 13, 1972 , near the nucleus of the elliptical galaxy NGC 7334 in Pegasus.The B and V light curves (Fig.3) are characteristic of a type I supernova,with a steep decline (0.090 magn./d) followed by a slower linear decline (0.014 mag./d).The B-V color index was negative near maximum, but rose to about + 1 during the early decline, then decreasing to nearly zero at mid November.The maximum was attained on August 21 , 1972, with B= 14.2

Assuming for this supernova at maximum an absolute magnitude -18.6 (Kowal, 1968), its distance and that of the parent galaxy fall between 31 and 36 Mpc , according that a correction for galactic absorption is applied or not. The galaxy very likely belongs to the same physical group as NGC 7631 , 7617, 7619, 7623, 7626 , all in the same area and with about the same distance.

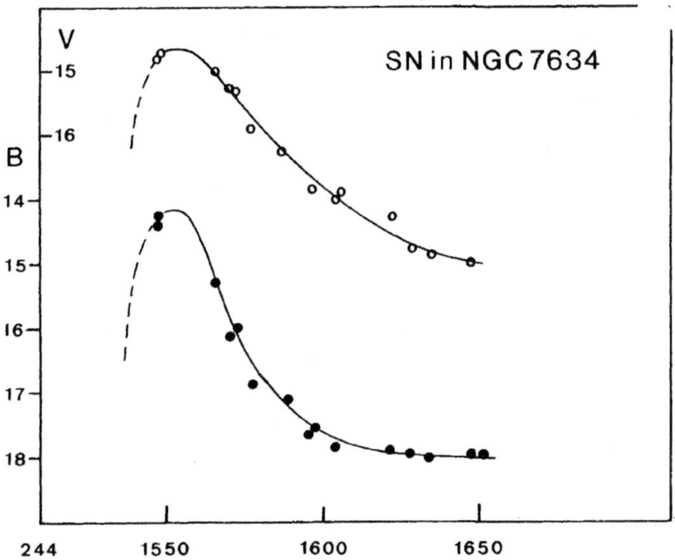

Fig.3 - B and V light curves of SN 1972-j

SN 1973-n in NGC 7495 - A supernova of magn.15.5 was discovered by Wild at a distance of 16" from the nucleus of the small spiral galaxy NGC 7495. The B and V light curves (Fig.4) are typical of SN-I . The maximum very likely occurred on August 29, 1973, with B= 15.4 ,then the supernova followed the normal pattern of photometric evolution. From the color curve it is clear that the star had a B-V color excess of about +0.6. Assuming for the supernova an absolute magnitude $M_B = -18.6$ at maximum , the corresponding distance is d= 21 Mpc.

The spectrum was also typical of a SN-I, but all the usual features were redward displaced of a quantity z = 0.014, corresponding to a symbolic velocity of 4200 km/s After correction for the solar motion we found, for the distance of the galaxy, d= 44.2 Mpc , somewhat different

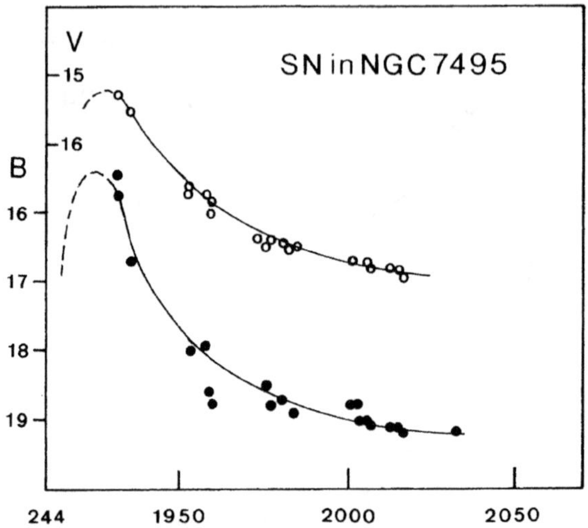

Fig.4 - B and V light curves of SN 1973-n

from the precedent value. Since both determinations can be affected by serious errors, we have adopted for the distance of the galaxy the mean value d = 32.6 Mpc.

SN 1973-r in NGC 3627 - This supernova was discovered at Asiago as a star of magn.14.5 in a spiral arm of the SBb galaxy NGC 3627 = M 66. The B and V light curves (Fig.5) are typical of a SN-II. The maximum (B=15) was attained on Dec 14-15, 1973. The color curve indicates that the supernova had a color excess E_{B-V} = 0.6, probably due to absorption within the parent galaxy. The absolute magnitude at maximum, M_B =-16.8 , was derived from the distance of the galaxy. A second value (-16.0) was obtained from the brightness of the plateau (de Vaucouleurs, 1974).

The spectra secured from December to March confirm that the supernova is type II.They show a strong continuum with superposed wide emission bands flanked by broad P Cyg absorptions. All features shift progressively towards the red, while the expansion velocities of the absorbing layers are decreasing from about 11000 km/s on Dec 28 to 7000 km/s on March 14.

There is no doubt about the presence of the Balmer lines in the spectrum, while the identification of the other lines is a more difficult problem. There are two alternatives: the first (Minkowski 1939, 1941) assumes that the emission features are due to He, N, O , at different stages of ionization, the second (Kirshner et al., 1973) that they are produced, in a low temperature envelope, by atoms of NaI, MgI, CaII, FeII. We have preferred the first alternative , tentatively attributing the emissions to HeI, HeII, NII.

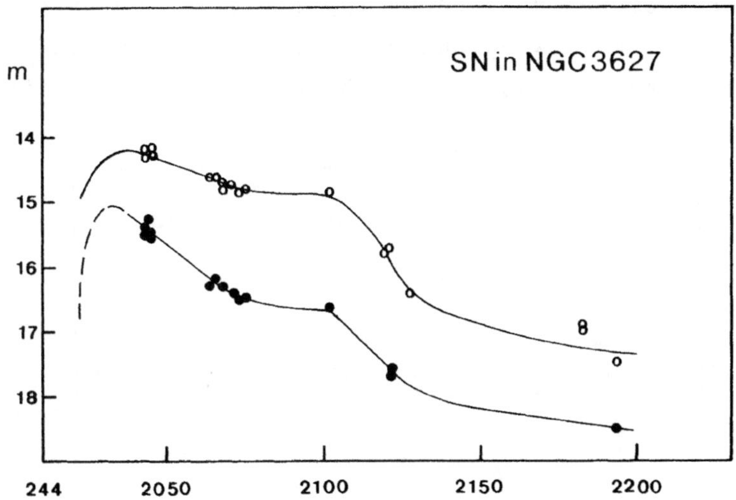

Fig.5 -B and V light curves of the type II supernova 1973-r

SN 1974-g in NGC 4414 - Burgat discovered on April 20 this supernova as a star of magn. 14 in the Sc galaxy NGC 4414. The B light curve (Fig.6) is typical of supernovae I.The maximum (B=12.5) was attained on April 30, 1974. The B-V color index increased from about zero at maximum, to + 1.4 forty days after maximum,then decreased to +0.75 at mid July.The best fit of the light curve of this supernova with the standard light curve of type I supernovae yields for the absolute magnitude at maximum the value $M_B = -18.4$ and for the distance d= 13.5 Mpc.

The spectrum of this supernova has been studied by Iye et al. (1975), by Patchett and Wood (1976) and by Ciatti and the writer. The wavelengths of absorption and emission features derived by the different observers agree fairly well. Following Mustel (1971), Branch and Patchett (1973) we have tried to attribute the absorption features

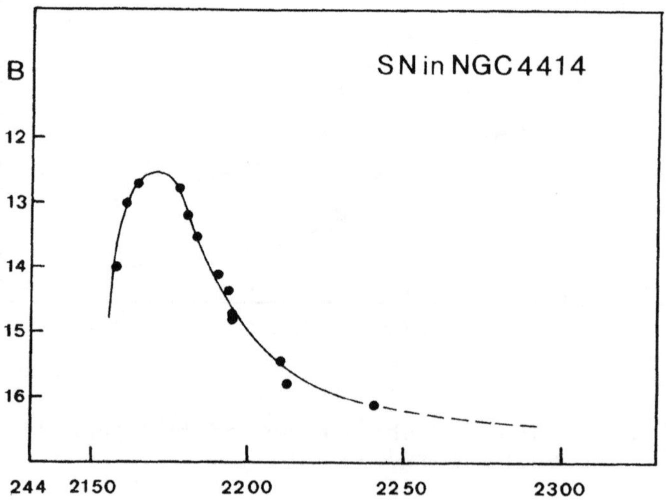

Fig.6 -B light curve of the SN 1974-g

to SiII (λλ 6347-61),NaI, ScII, FeII and H_β ,assuming that they are shortward displaced by a radial velocity of -8600 km/s due to rapid expansion of the ejected shells.However, some of the coincidences are very doubtful and the problem of the identification,in our opinion, is still widely open.

SN 1974-j in NGC 7343 - This supernova was discovered at Asiago at a distance of 18" from the nucleus of NGC 7343, a small SBc galaxy close to NGC 7331 and to the Stephan's Quintet. The B and V light curves (Fig.7) are typical of a type I supernova,with a relatively fast decline of 2.6 magnitudes in the first thirty-five days,followed by a slower decline (0.017 mag/d).The maximum was reached near October 13, 1974, with B magnitude nearly 16. Assuming for the supernova at maximum the absolute magnitude M_B = -18.6 the distance of the parent galaxy, after correction for galactic absorption, is d= 61.4 Mpc.

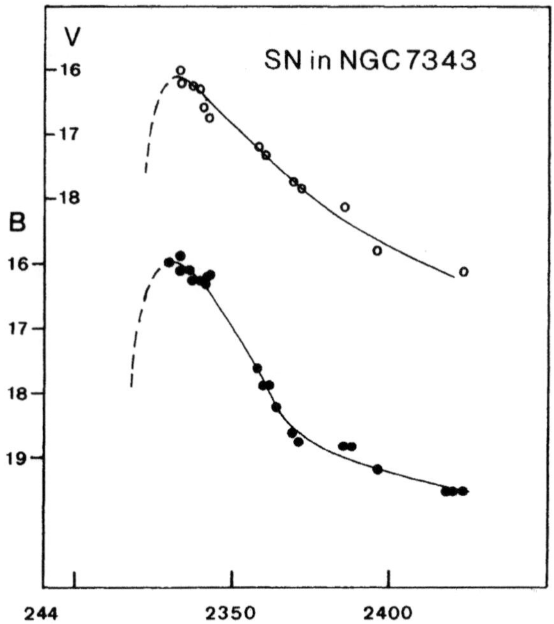

Fig.7 - B and V light curves of SN 1974-j

This value is confirmed by the analysis of the spectrum, which is typical of type I supernovae, but shows a redshift $z = 0.018$, corresponding to a galactocentric velocity of 5680 km/s and to a cosmological distance $d = 56.8$ Mpc.

The agreement of the two independent distance determination gives evidence that the redshift-distance relation in this region is not violated and that the distance of the galaxy corresponds to its redshift.Very likely the galaxy forms a physical group, at a distance of about 60 Mpc , with NGC 7335,the four high-redshift members of the Stephan's Quintet and some other galaxies in the same region.

REFERENCES

BARBON,R.,CIATTI,F.,ROSINO,L.:1973a,Astron.& Astrophys. 25,241.

BARBON,R.,CIATTI,F.,ROSINO,L.: 1973b,Mem.Soc.Astron.It.,44,65.

BARBON,R.,CIATTI,F.,ROSINO,L.: 1973c,Astron.& Astrophys.29,57.

BRANCH,D.,PATCHETT,B.: 1973,Montly Notices Roy.Astron.Soc. 161,71.

IYE,M.,KODAIRA,K.,KIKUCHI,S.,OHTANI,H.:1975,Publ.Astron. Soc.Jap.27,571.

KIRSHNER,R.P.,OKE,J.B.,PENSTON,M.V.,SEARLE,L.:1973,Astrophys.J. 135,303.

KOWAL,C.T.:1968,Astron.J. 73,1021.

MINKOWSKI,R.:1939,Astrophys.J.89,143.

MINKOWSKI,R.:1941, Publ.Astron.Soc.Pacific 53,224.

MUSTEL,E.R.: 1971, Soviet Astron. 15,1.

PATCHETT,B.,WOOD,R.:1976,Montly Notices Roy.Astron.Soc. 175,595.

ROSINO,L.: 1964, IAU Circular No.1868.

ROSINO,L.: 1971, Inf.Bull.Variable Stars No.515.

ROSINO,L.,di TULLIO,G.: 1974,Supernovae and Supernova Remnants,Ed.C.B. Cosmovici,Reidel.

VAUCOULEURS,G.de.: 1974,Supernovae and Supernova Remnants,Ed.C.B. Cosmovici,Reidel.

THE EARLY TYPE I SUPERNOVA LIGHT CURVE: THE EFFECT OF HYDROGEN ABUNDANCE

Gordon Lasher
IBM-T. J. Watson Research Center

Alan H. Karp
Dartmouth College

K. L. Chan
University of Calgary

Since the early days of Colgate and White (1965), most researchers working with numerical models of supernovae have specialized in either the inside or the outside parts of the supernova. This development is a reasonable one. The gravitational collapse of a dense core is completed in a second or two, but the passage of a shock wave through a stellar envelope may take half a day. To model the shock wave and the subsequent observable emission of radiation one only needs to know the energy, momentum and material deposited in envelope by the inner processes. This paper reports our most recent progress in modelling the early part of Type I light curves.

The rising portions of supernova light curves have two segments. The first very abrupt rise occurs when the shock wave erupts from the stellar surface. The subsequent rise is much more gradual as the radiation diffuses out of the expanding shell. In the ten solar mass model of Falk and Arnett (1976) the light level attained at shock breakout is nearly the maximum value. In still heavier models the light curve would occur during shock breakout. All or almost all observed Type II light curves show only a declining luminosity, no observations having been made during the period of increasing luminosity. The lack of such observations is to be expected for very heavy stars since the rising portion of the light curve is exceedingly brief.

On the other hand a recently published model for Type II supernovae by Chevalier (1976) having an envelope mass of 4.7 solar masses achieves a maximum luminosity well after the emergence of the shock wave. In this case the radiation escapes by diffusing out of the interior of the expanding shell as in our Type I models. It is possible that both points of view are correct, and that there are two subclasses of Type II supernovae which are presently lumped together. There does not, however, seem to be any

David N. Schramm (ed.), Supernovae. 13-20.

obvious correlation of this proposal with the finer classification of supernovae used by F. Zwicky (1965).

The light curves of Type I supernova indicate that they are a quite homogeneous class of objects (Barbon, Ciatti and Rosino 1973). Colgate and McKee (1969) proposed that the radiation emitted in the region of the maximum brightness was diffusing out of an expanding shell of gas. Kirshner et al. (1973) interpreted their spectrophotometric observations of the Type I supernova 1972e in terms of the emission of black body radiation.

These results and the encouragement of C. Gordon lead to our work on the very simple models reported in a recent paper by one of us (Lasher 1975). These models begin with a static sphere of gas representing a supergiant envelope. According to modern stellar structure computations most of this envelope has a nearly constant density (Bodenheimer and Ostriker, 1974). Into the center of the sphere of gas a quantity of thermal energy is deposited by the collapse or detonation of the stellar core.

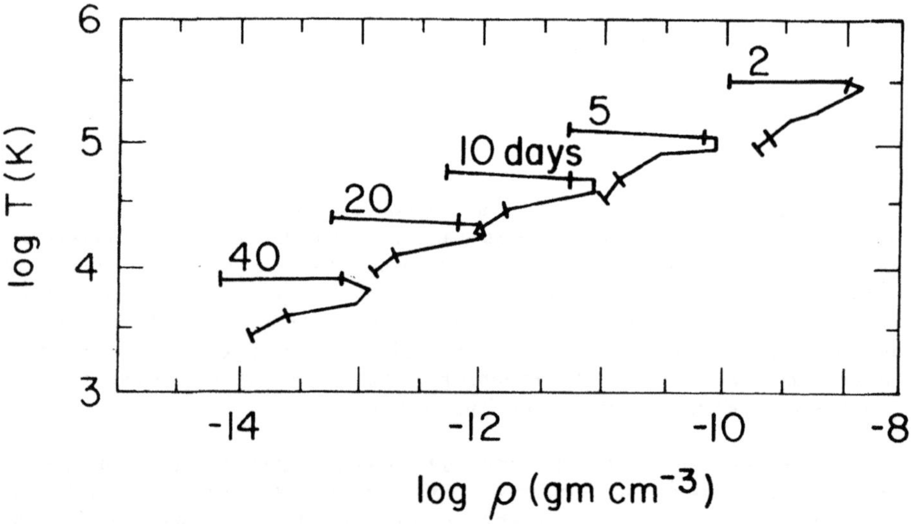

Figure 1.-The temperature-density traces of a typical constant opacity supernova model (Lasher 1975). The model parameters are an envelope mass of two solar masses, an energy of 10^{51} ergs and an initial density of 10^{-8} gm cm^{-3}. The labels on the lines give the time in days since the explosion.

The thermal energy deposited in the center of the envelope causes it to expand and generate a shock wave which reaches the surface in about half a day. The emergence of the shock wave causes the emission of a burst of hard ultraviolet and soft x-ray radiation. The visible luminosity at this time is, however, smaller than at later times because the radius of the shell is still comparatively small. The surface temperature of the shell falls rapidly following the emergence of the shock wave and the visible luminosity falls with it for a few hours.

The expansion continues, and then the visible luminosity increases again as the trapped radiation diffuses out of the thinning shell. After ten days or so the supply of radiant energy is depleted by the simultaneous processes of adiabatic expansion and emission, and the visible luminosity again decreases. The result is that the visible and photographic magnitudes show a broad maximum ten to twenty days after the core collapse initiated the explosion.

Figure 1 shows some temperature-density traces of one of the simple models during this expansion phase. The lines labelled by the time since the explosion connect the temperature-density points of the fifteen radial zones of the model. The tick marks locate the points corresponding to the innermost and outermost zones, and to the zones adjacent to them. The two innermost zones contain roughly half of the radiative energy; the supernova shell is a bubble of radiation which gradually leaks out of its containing skin of plasma. Up to ten days the central temperature varies as the cube root of the density, that is, at the adiabatic rate. Thereafter it deviates downward as the radiation escapes. This model yields the lower light curve of Figure 2.

The opacity of the shell as given by the Cox and Stewart (1970) tables is for the most part close to the Thompson scattering value for fully ionized matter. The exception occurs for the outside of the shell at late times when recombination of the electrons becomes important. In our original work we assumed that the opacity was constant with frequency and temperature. One result of the assumption of a constant opacity was that models having only fifteen radial zones proved adequate to predict light curves to within tenth of a magnitude. We fit the models to the observations by generating a net of models and linearly interpolating between the members of the net.

A total of three numbers determined these models, the mass and initial density of the stellar envelope and the energy deposited by the core collapse. Three observational numbers then served to determine the values of these input numbers. We chose to fit the model to the maximum blue magnitude, the length of time that the brightness remained within one magnitude of its maximum value, and the terminal velocity of the outside of the supernova shell.

The models which gave a good fit to the observations typically had an envelope of two solar masses, an initial density of 10^{-8} gm cm^{-3} and an energy of 10^{51} ergs. These numbers are only preliminary values for reasons that we will discuss, but the fact that they fell in a plausible range convinced us that this approach was worth further development. The agreement of the observed color curve of the blue magnitude versus time was also encouraging.

As we have already mentioned our first models assumed that the opacity of the supernova shell was simply that due to the Thompson scattering of the completely ionized gas. Aside from the obvious discrepancy due to recombination at low temperatures, this assumption is in agreement with published opacities for the temperature and density ranges of the models (Cox and Stewart 1970). One difficulty was apparent in this work however. Electron scattering hardly changes the frequency of the scattered photons and therefore does not maintain the radiation in thermal eguilibrium as it diffuses through the expanding shell. If the diffusing radiation is not kept in thermal equilibrium it emerges 'hotter' or 'bluer' than predicted by our models.

In our initial paper (Lasher 1975) it was argued that in an expanding shell there is an additional effect which tends to keep the radiation in thermal equilibrium and at the same time increases the effective opacity. This effect is simply that the photons travelling through an expanding object suffer a continuous redshifting with respect to the rest frame of the expanding gas. As a result of this differential doppler shift a typical photon has an increased chance of interacting with a discrete atomic transition. In a current preprint (Karp et al., 1976) we derive a formula for this effective opacity in expanding objects which we term the expansion opacity. It is valid in the interior of optically thick objects and gives the expansion opacity in terms of a sum over spectral lines. Calculations of the expansion opacity using the list of 260,000 lines given by Kurucz and Peytremann (1975) are reported in that preprint. These results show that the expansion opacity relevant to our models is typically two or three times the ordinary static opacity. We conclude that this expansion effect can be effective in keeping the radiation in the expanding shell in thermal equilibrium and should be included in precise models of Type I supernovae.

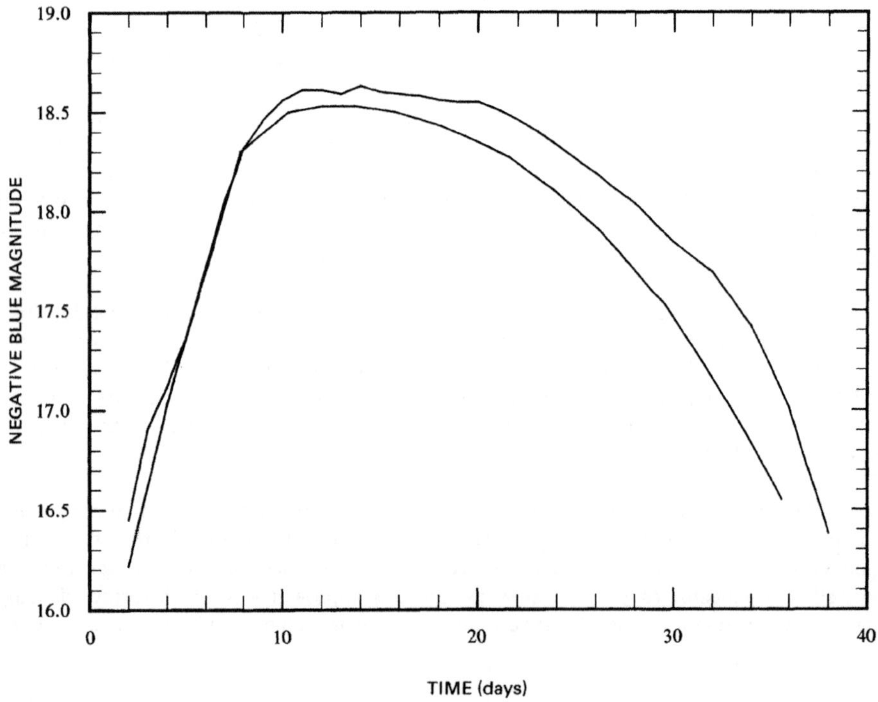

Figure 2.-The light curves of two models chosen to illustrate the effect of using the expansion opacity of Karp et al. (1976) instead of a constant opacity. The lower curve is from the constant opacity model and the upper one the expansion opacity model, the model parameters are the same as those of Figures 1 and 3.

The model light curves of Figure 2 show the effect of using the expansion opacity of Karp et al. (1976) in place of the constant opacity. The lower curve is for the fifteen zone model with the constant opacity equal to .2(1+X) where X, the weight fraction of hydrogen, was taken to be one tenth. The upper curve is for a fifty zone model using the Rosseland mean of the expansion opacity which includes not only the enhanced spectral line effects discussed in Karp et al., but also the reduction of opacity which accompanies the electronic recombination. The net effect is to lengthen the duration of the light curve by about three days.

The three light curves of Figure 3 are from models which use the Karp opacities and differ only in the assumed abundance of hydrogen. The abundances of the normal hydrogen model are cosmic abundancces with hydrogen, helium, and heavier elements having weight fractions X=0.78, Y=0.24 and Z=0.02. In the low hydrogen and zero hydrogen models the abundances are changed by converting enough hydrogen to helium to make the weight fraction X equal to 0.1 and 0 respectively.

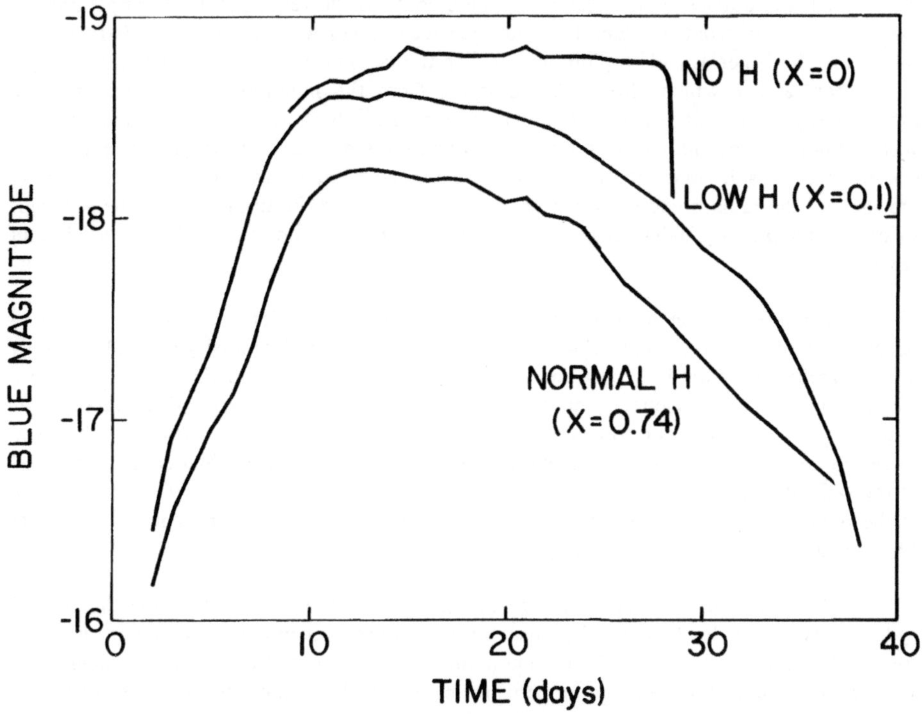

Figure 3.-The light curves, blue magnitude B versus time in days, predicted by three models having a weight fraction of hydrogen of 0, 0.1, and 0.78. All three models have an initial envelope mass of two solar masses, an initial density of 10^{-8} gm cm^{-3} and an explosive energy of 10^{51} erg.

The normal hydrogen light curve of Figure 3 is similar to those of the constant opacity models. One difference is that the final decrease is approximately a straight line whereas the light curves of our previous models are concave downward. The straight line decrease would give a somewhat better fit to the average light curve of Barbon et al. (1973), but such comparisons of detailed shape should be made with an individual light curve rather than the average light curve.

The low hydrogen light curve of Figure 3 has a higher maximum brightness than the normal hydrogen abundance because of the decreased electron density and therefore lower opacity associated with the lower hydrogen content. This curve is strongly concave downward after the maximum, and we believe that this is due to the recombination of helium which decreases the opacity at the lower temperatures.

The no hydrogen light curve of Figure 3 is radically different from the other two, and this difference is due to the effect of what is called a cooling or recombination wave. Grassberg, Immshennik and Nadyozhin (1971) first considered in some detail the effect of a recombination wave on supernova models containing significant amounts of hydrogen. When the outside of the expanding shell cools to the recombination temperature of hydrogen, a sharp front in which the hydrogen recombines with the electrons is formed at the surface and then travels rapidly inward. In the layers outside the front there is only a nearly neutral gas which allows the radiation from the neighborhood of the front to escape easily. This easy escape of radiation from the neighborhood of the recombination wave allows the recombination process to procede rapidly and causes the wave to move inward at a considerable velocity. When the recombination wave reaches the center of the shell all the radiant energy of the shell has escaped, and the brightness of the supernova drops precipitously unless there is some other structure inside the shell which will emit radiation.

In models which have some hydrogen the recombination wave is initiated at a temperature of roughly 5000K which is reached only after the luminosity has fallen two or three magnitudes below its maximum. It therefore has no effect on the fitting procedure adopted for the constant opacity models. It is possible that the passage of the recombination wave is responsible for what Zwicki (1965) called a protusion, a portion of the Type I light curve at the beginning of the exponential tail which is slightly higher than the level of the tail extrapolated to earlier times. In the model C of Chevalier (1976) the passage of the recombination wave through the envelope reveals the much denser mantle material which continues to emit radiant energy.

The zero hydrogen model has a recombination wave, but in this case the role of hydrogen is played by helium. Because the first ionization potential of helium is much greater than that of hydrogen the recombination wave is formed at a higher temperature, namely near 10,000 K, which happens to be reached around the time of maximum luminosity. The result seen in Figure 3 is that the luminosity remains near its maximum for several days because the recombination wave releases the radiant energy more quickly than the outward diffusion of the other two models. There is nothing inside the shell of our model. When the recombination wave reaches the center, the supernova disappears. The resultant light curve is in sharp disagreement with all observed Type I light curves. We therefore conclude that Type I supernovae shells cannot be composed of almost pure helium. The admixture of any more easily ionized element such as hydrogen will delay the recombination wave and give a light curve whose shape will agree with those observed.

The new results reported in figure 3 are obviously quite rough because of the coarseness of the zoning and the rather crude approximation adopted for the radiative transfer. Besides refining our models and obtaining a more precise fit to the observations; we also plan to improve our estimate of the soft x-ray flash (Lasher and Chan 1975) which is generated by the shock wave as it bursts out of the stellar surface. The observation of this flash would of course determine the answers to many of the questions about the nature of supernovae. Models of our type require that the initial star have very large radius. The x-ray pulse from such a large object is extended in time because of the delay suffered by the radiation in traveling from various parts of its surface. Xray pulses from more compact precursors would be shorter in time and consist of higher frequency radiation (Colgate 1974). The model C of Chevalier (1976) would yield an x-ray pulse similar to ours. Falk and Arnett's models, on the other hand, would generate a quite small, perhaps unobservable, x-ray pulses which would appear at the same time as the visible supernova. When Chevalier's model or ours apply, the x-ray pulse precedes the visible supernova, and its observation would provide an opportunity for many optical observations to be made of the same event from the time of its first appearance.

Models which start at much higher densities than ours do not give reasonably bright supernovae because the adiabatic cooling which accompanies the expansion of the shell cools it to low temperatures before it is large enough to radiate a significant amount of light. If there is energy injection by radioactive elements during the expansion, the effect of adiabatic expansion is circumvented. Such models have been recently considered by Nadyozhin and Utrobin (1976).

Our models indicate that Type I supernovae originate from quite massive stars. An envelope mass of two solar masses implies a total mass of more than three solar masses because the mass of the core must be included. Many astronomers believe that the occurence of Type I supernovae in elliptical galaxies requires a much smaller mass for the parent star since the normal ellipticals presumably contain only old stars. Tinsley (1976) has given a recent review of the relevant evidence. As noted there, however, this point of view has been challenged by three papers: Dallaporta (1973), Larson and Tinsley (1974) and Tammann (1974). Even if one denies the possibility of star formation in ellipticals, the work of Dellaporta indicates that most of the observed Type I may have massive precursors, while the presence of supernovae in elliptical galaxies is accounted for by a relatively rare subclass of Type I.

Aside from the construction of numerical models of supernovae, there is a more direct way of deriving information from the observations. The work of Kirshner and Kwan (1974 and 1975) shows that one can interpret the observations in terms of a moving photosphere. Similar work on Type I observations has been reported by Mustel and Chugai (1975), Branch and Patchett (1973) and Branch (1976). The latter two papers show that one can derive a value of the Hubble constant from the Type I observations which is in agreement with the accepted value. Our models can be regarded as a quantitative description of the physical processes by which the moving photosphere is created from the original star.

The authors are indebted to C. Gordon and E. E. Salpeter for many helpful discussions.

REFERENCES

Barbon, R., Ciatti, F. and Rosino, L. 1973, *Astr. and Ap.*, **25**, 241.
Bodenheimer, P. and Ostriker, J. P. 1974, *Ap. J.*, **191**, 465.
Branch, D. 1976, these proceedings.
Branch, D. and Patchett, B. 1973, Mon. Not. Roy. *Astr. Soc.* **161**, 71.
Chevalier, Roger A. 1976, *Ap. J.*, **207**, 872 and these proceedings.
Colgate, S. A. 1974, *Ap. J.* **187**, 321 and 333.
Colgate, S. A. and McKee, C. 1969, *Ap. J.*, **157**, 623.
Colgate, S. A. and White, R. H. 1966, *Ap. J.* **143**, 626.
Cox, J. P. and Stewart, J. N. 1970, *Ap. J. Suppl.*, **19**, 243 and 261.
Dallaporta, N. 1973, *Astr. and Ap.* **29**, 393.
Falk, S. W. and Arnett, W. D. 1973, *Ap. J.(Letters)*, **180**, L65.
Falk, S. W. and Arnett, W. D. 1976, *Orange Aid Reprints* 454 and 455.
Grassberg, E. K., Imshennik, V. S. and Nadyozhin, D. K. 1971, *Ap. and Space Science*, **10**, 28.
Kirshner, R. P. and Kwan J. 1974, *Ap. J.*, **193**, 27.
Kirshner, R. P. and Kwan J. 1975, *Ap. J.*, **197**, 415.
Kirshner, R. P., Willner, S. P., Becklin, E. E., Neugebuer G, and Oke, J. B. 1973, *Ap. J.(Letters)*, **180**, L97.
Karp, A. H., Lasher, G., Chan, K. L. and Salpeter, E. E. 1976, preprint.
Kurucz, R. L. and Peytremann, E. 1975, Smithsonian Ap. Ob., Special Report 362.
Lasher, G. 1975, *Ap. J.*, **201**, 194
Lasher, G. and Chan, K. L. 1975, *Bull. Am. Astr. Soc.*, **7**, 505.
Larson, R. B. and Tinsley, B. M. 1974, *Ap. J.* **192**, 293.
Mustel, E. R. and Chugai, N. N. 1975, *Astron. Zh.*, **52**, 673 (English Trans., Sov. Astron., **19**, 411).
Nadyozhin, D. K. and Utrobin, V. P. 1976, two preprints.
Tammann, G. A. 1974, in *Supernovae and Supernova Remnants*, ed. Cosmovici, C. B. (D. Reidel) p. 155.
Tinsley, B. M. 1975 *Publ. Astr. Soc. Pacific* **87**, 837 and these proceedings.
Zwicki, F. 1965, in *Stellar Structure*, ed. Aller, L. H. and McLaughlin, D. B. (Univ. of Chicago Press), p.367.

SUPERNOVAE AND THE VALUE OF THE HUBBLE CONSTANT

David Branch
University of Oklahoma

1. INTRODUCTION

The Sandage and Tammann (1975) value of the Hubble constant, adjusted to the new distance to the Hyades, is 50 ± 5 km s^{-1} Mpc^{-1}, where the quoted uncertainty refers to the internal root-mean-square error but not the systematic errors, which are difficult to assess. On the other hand both de Vaucouleurs and van den Bergh, also using classical methods, continue to prefer values nearer to 100.

The thermal interpretation of the light emitted by supernovae during their brightest optical phases has opened up a new, independent approach to the value of H_o, by way of the Baade - Wesselink method. Branch and Patchett (1973) applied the method to the rising part of a composite light curve for Type I supernovae and found $H_o = 40^{+25}_{-13}$. Kirshner and Kwan (1974) applied the method to Type II supernovae in M101 and NGC 1058; the derived distances, combined with the assumption that the Hubble law holds strictly for these two relatively nearby galaxies, gave 60 ± 15. The agreement, within the quoted errors, of the two preliminary supernova results and the Sandage and Tammann distance scale is encouraging, and the importance of refining the supernova method can hardly be overemphasized.

At present the use of Type I supernovae rather than Type II has two distinct advantages, both having to do with availability of data. First, Type I supernovae have been found to be remarkably homogeneous in their photometric and spectroscopic properties, so the Baade - Wesselink method can be applied to composite light and color curves. The use of composite data for the more individualistic Type II's would be harder to justify. Second, if we want to avoid the controversy regarding the possible kinematic effects of the local supercluster (de Vaucouleurs 1976) and to have instead a genuinely cosmic value for H_o, we must use Type I supernovae. There is very little available data on Type II supernovae in galaxies well outside the local supercluster.

I will discuss a simple semi-empirical model of the envelopes of Type I supernovae at their very brightest phases, with emphasis on the use of the model

David N. Schramm (ed.), Supernovae. 21-28.

to obtain a good cosmic value for H_o.

2. THE MODEL

The envelope of a Type I supernova near the time of maximum brightness is assumed to be spherically symmetric and expanding differentially with velocity proportional to distance from the center. Each mass element coasts at constant velocity. The observed homogeneity supports the assumption of spherical symmetry. The differential expansion with $v \propto r$ would be the natural consequence of the outward propogation of an accelerating shock wave shortly after outburst, followed by a phase of free expansion. A well defined, cooling photosphere is assumed to emit a blackbody continuum, and outside the photosphere spectral lines are formed by scattering. The absolute magnitude and the color index are determined by the radius and temperature of the photosphere

$$M_B(t) = 53.6 - 5 \log R(t) + 35300/T(t) \quad (1)$$

$$B\text{-}V(t) = 6900/T(t) - 0.56 \quad (2)$$

where the numerical constants are those found by Arp (1961) for blackbodies. It follows that the absolute magnitude at the time of maximum brightness (t_o) is given by

$$M_B(t_o) = 53.6 - 5 \log R(t_o) + 5.1\,[B\text{-}V(t_o) + 0.56]. \quad (3)$$

The color at maximum brightness is observable, but the value of the radius of the photosphere remains to be determined. The change of the radius of the photosphere with respect to its value at maximum brightness, i.e., the radius curve, can be found directly from the observed changes in magnitude and color with respect to their values at maximum brightness

$$\log R(t)/R(t_o) = -0.2\,\Delta M_B(t) + 1.02\,\Delta B\text{-}V(t). \quad (4)$$

This equation, applied to the composite light and color curves published by Barbon, Ciatti, and Rosino (1973) gives an important result: over an interval of about 20 days centered on maximum brightness, the radius curve is linear.* During this interval the photosphere remains fixed in the material and expands at constant velocity.

---

* This and some other aspects of the model are discussed and illustrated in more detail in a paper which will appear in the Monthly Notices of the Royal Astronomical Society.

The linearity of the radius curve leads to a conclusion regarding the density structure of the envelope, near the photosphere. Suppose that the temperature and density in the envelope are expressed in terms of powers of the distance from the center and the time, and that the opacity is expressed in terms of powers of the temperature and density. It follows that the radius of the photosphere varies as

$$R(t)/R(t_o) = (t/t_o)^x \quad (5)$$

where x is a constant, whose value is determined by the temperature and density structure and the opacity law. The linearity of the radius curve and the homogeneity require that in each Type I supernova the temperature and density structure and the opacity law are just nicely arranged such that $x \simeq 1$. In the general case this would be rather coincidental, but if the density gradient at the photosphere is sufficiently steep the value of x approaches unity for a fairly wide range of temperature structures and opacity laws. Physically, it is simply that it is hard for the photosphere to propogate up or down a steep density gradient; it tends to remain fixed in the material. In the special case of a constant opacity, independent of temperature and density, and an r^{-n} density structure, the expression for x is $(n-3)/(n-1)$. So in this context $n = 10$, for example, represents a steep density gradient but $n = 4$ does not.

The value of the radius of the photosphere at maximum brightness is obtained by multiplying the reciprocal of the slope of the radius curve (16.4 ± 1.0 days) by the velocity of the photosphere. The velocity must be obtained from the spectral lines.

3. SPECTRAL LINES AND THE VELOCITY OF THE PHOTOSPHERE

Kirshner and Kwan (1974) have argued that in an envelope expanding with $v \propto r$, the velocity of the photosphere can be found from the absorption minimum of an optically thin line or from the "red edge of the absorption trough" of a stronger line. However, Kirshner and Kwan were not considering steep density gradients. Scattering profiles calculated according to the escape-probability method as developed by Castor (1970) show that in strong density gradients the velocity of the photosphere is given by the wavelength of the absorption minimum, for a fairly wide range of optical depths. The red edge of the absorption trough of a line formed in a strong density gradient is too vaguely defined to be useful, and the displacement to short wavelengths of a weak line is less than that corresponding to the velocity of the photosphere. The basic assumption underlying Castor's method is that "there is little physical change in the medium across the distance in which the Doppler displacement at the frequency of the line changes by an amount equal to the width of the line absorption profile". In the present case this is just the condition that the ratio of the velocity of the photosphere to the thermal velocity be much greater than n. This condition is very well satisfied.

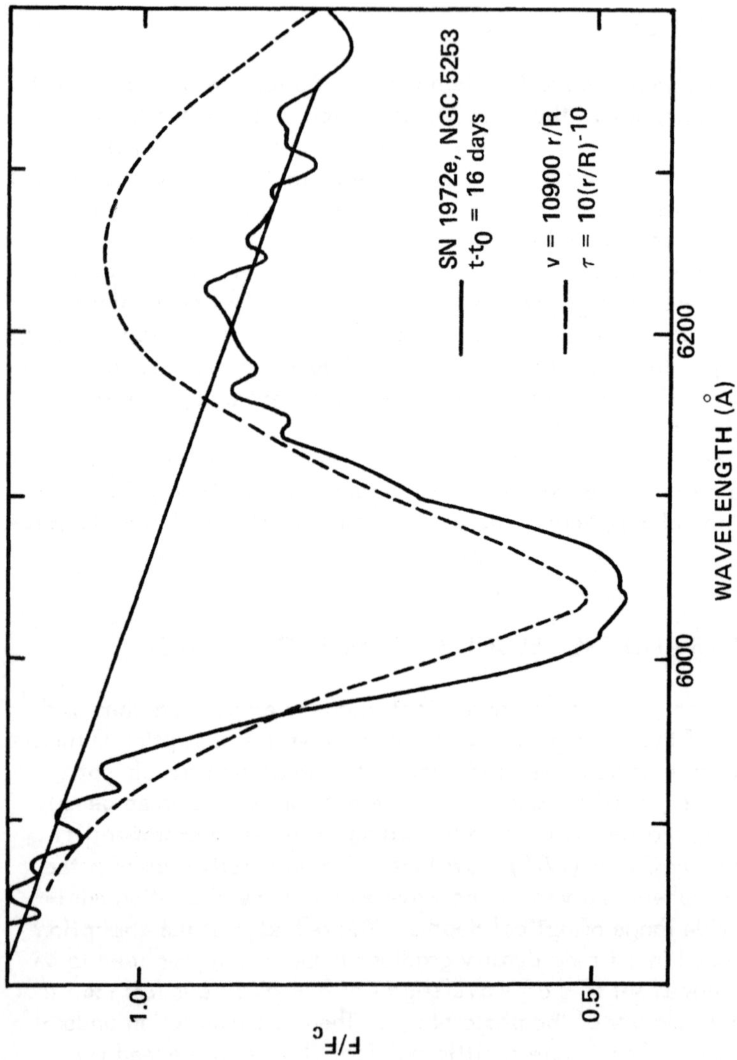

Fig. 1. - Comparison of a calculated and an observed line profile.

A comparison of an observed and a calculated profile is shown in Figure 1. The observed profile is from a photoelectric scan of SN 1972e in NGC 5253, at 8 Å resolution, made by R.G. Tull using the 2.7 - m telescope of the McDonald Observatory of the University of Texas. The supernova was about 16 days after maximum brightness. The absorption feature shown is characteristic of Type I supernovae until 25 or 30 days after maximum brightness. It is the only feature in the visible part of the spectrum which appears to be free from blending, and it does not drift in wavelength. The calculated profile is for an optical depth at the photosphere of 10, an r^{-10} density law, and a photospheric velocity of 10,900 km s^{-1} (see below). No attempt has been made to find the best fitting parameters; the purpose of the comparison is only to show that the calculated profiles can adequately reproduce the observed depths, widths, and displacements of the absorption components. The line identification adopted below is with a doublet of separation 24 Å. If this were taken into account the calculated profile would be broadened at the bottom, in better agreement with the observed profile. The calculated profile does show a noticeable emission component which is not observed.

To determine the velocity of the photosphere we need to use an unblended, moderately strong line. The feature shown in Figure 1 is suitable, provided that it is correctly identified. It has been identified with the 6347, 6371 Å doublet of Si II by Pskovskii (1969), Mustel (1971), and Branch and Patchett (1973). The mean wavelength of the absorption feature in spectra of 11 Type I supernovae is 6125 Å; the weighted rest wavelength of the doublet, 6355 Å, then corresponds to a velocity of 10,900 km s^{-1}. This is in accord with other line identifications (Ca II, Fe II, Na I) which have been suggested. Gordon (1972) identifies the feature with Hα, but then the corresponding velocity is 20,000 km s^{-1}.

If the density gradient at the photosphere is steep, the lines will tend to be formed near the photosphere. In this case the relative LTE line strengths at the photosphere should give a rough idea of which lines are to be expected. Calculations based on solar abundances and a simple model for the supernova envelope 15 days after maximum brightness predict that only H and Ca II lines would be strong, against the electron scattering opacity. Fe II lines would be marginally present. This may not be too far from the truth for Type II supernovae: strong lines of H and Ca II and weak lines of Fe II have been identified. (Patchett and Branch 1972, Kirshner, Oke, Penston, and Searle 1973, Kirshner and Kwan 1975). However, the most conspicuous differences between Type I and Type II spectra near maximum brightness are that the Balmer lines are not obviously present in Type I (excepting Gordon's identification of Hα), and that Type I contain more weak features than Type II. Calculations carried out on the assumption that hydrogen is sufficiently underabundant as to be a negligible source of free electrons predict that lines of Ca II, Fe II, Si II, and Mg II should be strong. The presence of Ca II lines is undisputed (?). Fe II lines are visible on slit spectra according to Pskovskii, Mustel, and Branch and Patchett, even though they are not so obvious on the lower resolution scans published by Kirshner, Oke, Penston, and Searle. The Mg II 4481 Å line falls among a group of

Fe II lines. These calculations provide some further support for the Fe II identification for the 6125 Å feature. The Si II doublet is expected to be present, on the basis of a simple model, and a feature does appear at the appropriate wavelength.

4. THE ABSOLUTE MAGNITUDE AND THE VALUE OF H_o

The adopted values for the velocity of the photosphere, 10,900 ± 700 km s^{-1}, and the reciprocal of the slope of the radius curve, 16.4 ± 1.0 days, give $R(t_o) = 1.54 \pm 0.14 \times 10^{15}$ cm. Combined with $B-V\ (t_o) = -0.15 \pm 0.05$, this gives the absolute magnitude of Type I supernovae at maximum brightness: $M_B\ (t_o) = -20.25 \pm 0.31$. The value of H_o is found from

$$\log H_o = 5 + (M_B(t_o) - C_B)/5 \tag{6}$$

where C_B satisfies the relation

$$B(t_o) = C_B + 5 \log cz \tag{7}$$

$B(t_o)$ is the apparent magnitude of a Type I supernova at maximum brightness and z is the redshift of the associated galaxy. The value of C_B, based on 9 supernovae in galaxies having $cz > 3000$ km s^{-1}, is -3.7 ± 0.3 (Branch 1974). Then

$$H_o = 49 \pm 9 \text{ km s}^{-1} \text{ Mpc}^{-1}. \tag{8}$$

This result supersedes the one given by Branch and Patchett (1973).

5. A COMMENT ON A RECENT ANALYSIS

An apparent failure of the supernova method has been reported by Kirshner, Arp, and Dunlap (1976). The method was applied to a 3-day interval during the rise to maximum brightness of SN 1975a in NGC 2207. The temperatures at the beginning and the end of the 3-day interval, based on 0.3 - 1.05 micron scans, were indistinguishable, so the observed change in brightness was attributed solely to a change of the radius. For $H_o = 50$ the distance should be 49 Mpc, but the derived distance was 187 Mpc. Kirshner et al. concluded that a longer time interval is required. In the form of the supernova method used here, the effect of a temperature change enters through the change in B - V over the interval in question. From Table 1 of Kirshner et al. the B - V change was 0.13. The method as used here then gives a distance of 40 Mpc.

The difference between 40 and 187 Mpc comes entirely from the difference in the ways that the temperature change is treated. The B - V curve for Type I supernovae is smooth and the line spectrum changes rather slowly, so I think it is better to use the accurately measurable (and reddening independent) change in B - V. Fitting the spectral scans by eye to blackbody curves is not sufficiently accurate for this purpose, despite the broader wavelength range.

6. ERRORS AND PROSPECTS

The error quoted above is the internal root mean square error. It originates more or less equally from the 4 parameters which enter into the final result: the reciprocal of the slope of the radius curve, the velocity of the photosphere, $B - V(t_o)$, and C_B. In principle it would be possible to reduce the error by making further observations, but at some point the accuracy would be limited by intrinsic differences between Type I supernovae. Anyway, we all know that the internal rms error can be misleading - the insidious systematic errors are the ones that really count. As far as the determination of $M_B(t_o)$ is concerned, the major question is the extent to which the adopted blackbody relations, equations (1) and (2), actually apply. However, the largest single uncertainty in the present result appears to be in the straightforward astronomical problem of determining the value of C_B. The present value is based only on relatively remote supernovae, in order to avoid the local supercluster. If C_B were based instead on all Type I supernovae, regardless of distance, its value would be −3.15 (Barbon, Capaccioli, and Ciatti 1975) and the value of H_o would come out to be 38. Therefore, although the formal result for H_o is 49, the best value may be in the range 38 to 49. If the model on which this analysis is based is essentially correct, it is not easy to see how to reconcile Type I supernovae with $H_o = 100$.

The precise value of the result reported today does not seem very important. What is important is the argument that the supernova method appears to be a valid one, capable eventually of precision. When sufficient data become available it will be interesting to apply the method to individual supernovae of all types.

A great deal of information about supernova envelopes is contained in scans such as the one shown in Figure 1. The scans obtained by Tull are the first ones of sufficiently high resolution for detailed studies of the lines. A comparison of synthetic spectra and a series of such scans is in progress, in the hope of testing and refining the model.

7. ACKNOWLEDGEMENTS

I wish to thank Dr. R.G. Tull for providing data in advance of publication. Research and travel has been supported in part by the United States National Science Foundation.

REFERENCES

Arp, H.C., 1961, Astrophys. J. 133, 874.
Barbon, R., Capaccioli, M., and Ciatti, F., Astr. Astrophys. 44, 267.
Barbon, R., Ciatti, F., and Rosino, L., 1973, Astr. Astrophys. 25, 241.
Branch, D., 1974, in C.B. Cosmovici (ed.), Supernovae and Supernova Remnants, D. Reidel Publishing Company, p. 209.
Branch, D., and Patchett, B., 1973, Mon. Not. R. astr. Soc. 161, 71.
Castor, J.I., 1970, Mon. Not. R. astr. Soc. 149, 111.
de Vaucouleurs, G., 1976, Astrophys. J. 205, 13.
Gordon, C., 1972, Astr. Astrophys. 20, 79 and 87.
Kirshner, R.P., Arp, H.C., and Dunlap, J.R., 1976, Astrophys. J. 207, 44.
Kirshner, R.P., and Kwan, J., 1974, Astrophys. J. 193, 27.
Kirshner, R.P., and Kwan, J., 1975, Astrophys. J. 197, 415.
Kirshner, R.P., Oke, J.B., Penston, M.V., and Searle, L., 1973, Astrophys. J. 185, 303.
Mustel, E.R., 1971, Sov. Astr. 15, 1.
Patchett, B., and Branch, D., 1972, Mon. Not. R. astr. Soc. 158, 375.
Pskovskii, Y.P., 1969, Sov. Astr. 12, 750.
Sandage, A., and Tammann, G.A., 1975, Astrophys. J. 197, 265.

X-RAYS FROM SUPERNOVA REMNANTS

J. L. Culhane
Mullard Space Science Laboratory,
University College London,
Holmbury St. Mary, Dorking, Surrey, England.

1. INTRODUCTION

Supernova remnants were the first class of X-ray emitting objects to be positively identified. Indeed they were the only sources of X-ray emission to be predicted as such before the beginning of observational X-ray astronomy in 1962. While the use of X-ray detection systems on sounding rockets lead to a number of advances in the study of remnants, spectacular progress has followed the availability of satellite borne instruments.

In this review we will briefly discuss the evolution of supernova remnants with special reference to the production of X-radiation and to the effect of inhomogeneities in the interstellar medium. The X-ray observations of the Crab Nebula, the younger ($t < 1000$ years) and the older remnants will then be reviewed in the light of our present understanding of the evolutionary phases.

2. X-RAY PRODUCTION IN SUPERNOVA REMNANTS

It is apparent in the case of the Crab Nebula that most of the X-ray emission is due to synchrotron radiation from relativistic electrons. However for all other X-ray emitting remnants, there is strong evidence that emission from hot gas is mainly responsible for the observed fluxes. Following the supernova explosion, a shock wave propagates into the surrounding interstellar medium (ISM). The kinetic energy of the shock wave is transformed into thermal energy in the gas so that the heated material radiates in the X-ray spectral range. It is usual to identify four phases in the expansion of the shock wave into the ISM.

I. Free Expansion Phase

In this phase

David N. Schramm (ed.), Supernovae. 29-51.

$$M_E > \frac{4}{3} \pi r_s^3 (t) \; n_I \tag{1}$$
$$> M_S$$

where M_E is the mass of material ejected in the explosion, M_S is the mass of interstellar material swept up by the shock at time t, $r_s(t)$ is the shock radius at time t and n_I is the density of the local ISM. Phase I ends when M_S has become several times greater than M_E or when $r(t) \simeq 2(M_E / n_I)^{\frac{1}{3}}$ Pc (equation 1). The interstellar matter entering the shock will be heated to a temperature

$$T = \frac{3m_p \mu V_s^2}{16k} = 2.10^{-9} V_o^2 \tag{2}$$

(Landau and Libshitz (1959) where $V_s \simeq 4\, V_o/3$ is the shock wave velocity, V_o is the expansion velocity of the shell of ejected material and μ is the average molecular weight for the ISM. The heated gas forms a spherical shell of thickness $r_s/12$. For gas temperatures between 10^7 and 10^8 ^{o}K, the luminosity $Lx \simeq 5.10^{33}\, n_I^2\, r_s^3$ erg/sec where r_s is in parsec (Pc) (Gorenstein, Harnden and Tucker (1975)).

The interaction of the shock wave with a dense cloud has been considered by Sgro (1975) and by McKee and Cowie (1975). The existence of clouds provides a source of material at higher than average density and, in young (phase I) remnants, such regions may have enhanced X-ray brightness. Finally McKee (1974) and Gull (1975) have pointed out that the deceleration of the ejecta by the surrounding ISM causes a shock wave (the "reverse shock") to propagate inwards through the ejected material at a lower velocity than that of the main shock. Since $M_E > M_S$ in phase I, a substantial X-ray flux can result from the heating of this material although its temperature will be lower than that of the interstellar gas heated by the main shock.

The calculation of the X-ray spectrum emitted in phase I is difficult since the heated gas probably differs considerably from a steady state "coronal" plasma. The fact that the ionisation and recombination times in a gas with $n_I \simeq 1 cm^{-3}$ are comparable to the expansion time scale of the shell could lead to enhancements in certain line intensities relative to the equilibrium coronal values (Kafatos and Tucker (1972)). Conversely if account is taken of the photoionisation of the ISM by X-rays from the original explosion, it is possible that some line intensities may be reduced (Chevalier (1975)). At the present time the only detailed calculations of X-ray spectra available refer to steady state coronal conditions (e.g. Tucker and Koren (1971), Mewe (1972)). Since emission lines have now been identified in the spectra of at least three young remnants, more appropriate calculations of the spectra are urgently needed

II. Adiabatic Phase

In this phase, $M_S > M_E$ but radiative losses are small compared to the initial energy of the explosion and so energy is conserved. It is therefore possible to employ the similarity solutions for the motion of a blast wave into a medium (Taylor (1950), Sedov (1959)). For a gas of specific heat ratio $\gamma = 5/3$, the value of the shock radius as a function of time is

$$r_s = 2.15.10^{-11} \frac{E_o}{n_I}^{\frac{1}{5}} t^{\frac{2}{5}} \text{ Pc} \tag{3}$$

where E_o is the total released energy in ergs and t is in units of years. The velocity of the shock

$$V_s = 8.4.10^{-6} \frac{E_o}{n_I}^{\frac{1}{5}} t^{-\frac{3}{5}} \text{ Km sec}^{-1} \tag{4}$$

while the temperature of the gas immediately behind the shock is given by

$$\begin{aligned} T_S &= 11.3 V_S^2 \\ &= 8.1.10^{10} \frac{E_o}{n_I}^{\frac{2}{5}} t^{\frac{6}{5}} \, {}^oK \end{aligned} \tag{5}$$

The density behind the shock $n_S = 4n_I$. The shock heated material is still confined in a thin shell (Δr) and thus the X-ray luminosity in an energy interval $E \pm \frac{\Delta E}{2}$ is

$$\begin{aligned} L_x (E) &= 4\pi r_s^2 \Delta r n_I^2 P(\Delta E,T) \\ &\simeq 6.10^{55} n_I^2 D^3 P(\Delta E,T) \end{aligned} \tag{6}$$

where D is the remnant diameter in Pc and $P(\Delta E,T)$ is the X-ray volume emissivity in a given energy interval.

In analysing X-ray observations of the Cygnus Loop, Stevens (1973) found it convenient to consider the expansion of the shock wave into a density gradient of the form

$$n = n_I \left(\frac{r}{r_S}\right)^{n'} \tag{7}$$

where n is the gas density as a function of r and r_S is the shock radius. Charles (1976) has incorporated this equation into the Sedov similarity solutions and, if $n' = -W$, finds

$$r_S = 11.6 \left(\frac{E_o}{\alpha n_I}\right)^{\frac{1}{5}} t^{\frac{2}{5}} \quad Pc \tag{8}$$

and

$$V_s = 98 \left(\frac{2}{5-W}\right) \cdot \frac{r_s}{t} \quad Km\ sec^{-1} \tag{9}$$

with

$$\alpha = \frac{12.328 - 1.244\ ln\ (1 - W)}{(5 - W)^2} \tag{10}$$

to be the appropriate expressions for shock radius and velocity in the presence of a density gradient.

In the adiabatic phase, the gas temperature increases towards the centre of the remnant. Solinger et al (1975) have suggested that the predicted large temperature gradients can not be maintained and have therefore derived a self-similar solution for an isothermal blast wave. However neither the X-ray structural or spectral data are yet of sufficient quality to permit a choice between these two approaches. (Charles (1976)).

III. Radiative Phase

According to Cox (1972), radiative loss became important when the shock radius has reached a value

$$r_c \simeq 3.1.10^{-13} \; \frac{E_o}{n_I}^{\frac{5}{17}} \; n_I^{\frac{-2}{7}} \tag{11}$$

For values of radius greater than r_C, the shell temperatures will be in the range 10^4 - 10^5 °K. While there may be some X-ray emission from residual hot gas near the centre of the remnant, L_X will be very much less than it was during the adiabatic phase.

IV. Disappearance Phase

When the expansion velocity becomes less than the thermal velocity of the interstellar gas (i.e. $\simeq$ 10 Km sec^{-1}), the remnant loses its identity and is no longer visible.

We will now discuss the application of these ideas to recent observations of the X-ray emission from supernova remnants.

3. THE CRAB NEBULA

In discussing this object, we will concentrate on the synchrotron emission from the nebula and on the possible existence of a thermal X-ray source outside the synchrotron source. Results on the nature of the pulsar itself will not be reviewed.

3.1. Synchrotron Emission from the Nebula

As was remarked above, the bulk of the X-ray emission from this object is due to synchrotron radiation from relativistic electrons produced in the pulsar which interact in the magnetic fields of the nebula. The synchrotron nature of the radiation has long been established by radio and optical observations and was further confirmed by the detection of polarised X-ray emission by Novick et al (1972). More detailed polarisation measurements have recently been reported by Weisskopf et al (1976) using a graphite crystal Bragg polarimeter on the OSO-8 spacecraft.

Observations of the X-ray structure of the Crab Nebula were obtained during the series of lunar occultations of this source that took place between August 13th and December 28th, 1974.

Energy Range	Angle between pulsar and X-ray centroid (arc sec)	Size (arc sec)	Position angle of observation (deg)	Reference
1.5-7.5 keV	20 ± 8 (W)	73 ± 38	236	Davison et al (1975) Satellite observations
		71^{+15}_{-8}	275	
20-150 keV	10 ± 4 (NW)	24 ± 7 49 ± 7	130 244	Ricker et al (1975) Balloon observation
15-150 keV	8 ± 5 (W)	110± 25	263	Staubert et al (1975) Rocket observation
1.5-17.5keV	~20 (w)	42 ± 2	255	Wolff et al (1975) Rocket observation
0.6- 23 keV	12.8±0.3 (W)	60.5±1.5	300	Kestenbaum et al (1975) Rocket observation
0.5- 16 keV	~8 (W)	100	278	Palmieri et al (1975) Rocket observation
20-150 keV	6 ± 4 (W)	34^{+17}_{-14}	102	Fukada et al (1975) Balloon observation

Table I. Results of Crab Nebula Lunar Occultation Studies in 1974.

The results of the 1974 occultations are summarised in Table I. The size estimates have been made on the basis of different distribution models (e.g. Gaussian, parabolic, rectangular etc.) and the original references should be consulted for details. The kind of data obtained from a rocket occultation observation are illustrated in Figure 1 from the work of Palmieri. Data of comparable statistical quality have been obtained by the Columbia group (Wolff et al (1975), Kestenbaum et al (1975)) at energies below 20 keV.

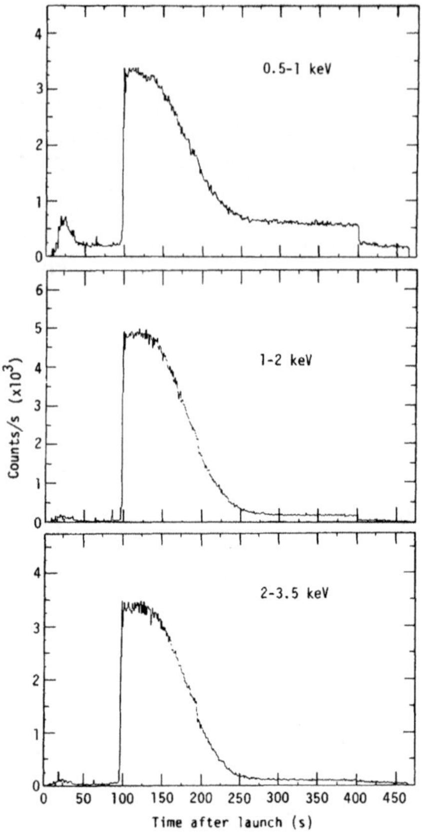

Figure 1. X-ray counting plotted against time during the Livermore occultation. The low energy flux excess following immersion of the nebula can be seen in the upper two plots.

The lunar occultations have led to a number of observational conclusions which are summarised below :-

1. The centroid of nebular X-ray emission is offset from the pulsar (Table I).
2. The offset angle decreases with photon energy (Toor and Seward (1976)). (See Fig. 2).
3. The size of the X-ray emitting region decreases with photon energy (Ku et al (1976), Toor and Seward (1976)). (See Fig. 2).
4. The X-ray source is asymetric with its long direction parallel to the nebular magnetic field vector (Ricker et al (1975), Toor et al 1976)). (See Fig. 3).
5. High energy ($20<E<160$keV) X-ray emission appears to be associated with the optical wisps (Ricker et al (1975), Staubert et al (1975)).

6. The X-ray surface brightness distribution favours constant emissivity throughout the radiating volume. An inverse square law dependance of emissivity on distance from the pulsar is ruled out (Kestenbaum et al (1975), Palmieri et al (1975)).

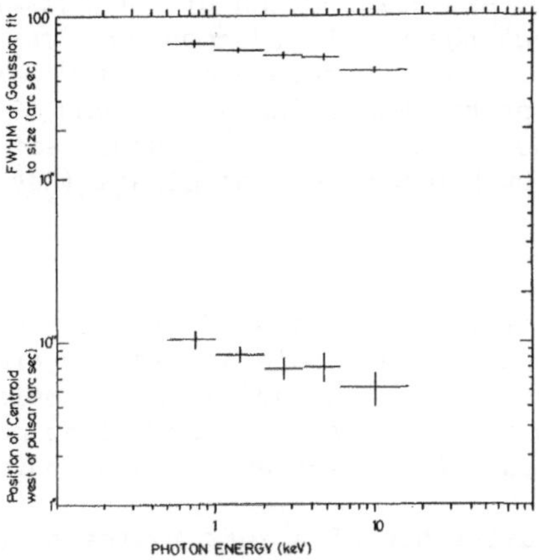

Figure 2. Results from the Livermore Laboratory's Crab occultation observation (see also Figure 1) which show the decrease in size and X-ray centroid offset as a function of photon energy.

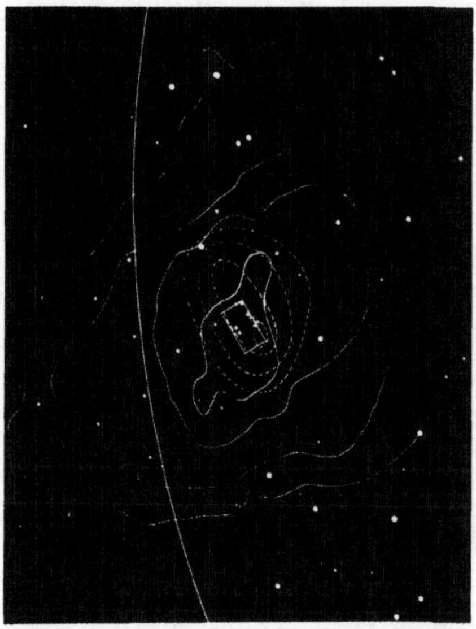

Figure 3. Crab occultation results. Low energy X-ray contours shown as dotted lines (Toor et al (1976). Solid rectangle shows location of hard (E> 20keV) source (Ricker et al (1975)). Solid line to left of map represents position of lunar limb at end of occultation. Solid contour lines represent radio source.

It would appear from the observed dependence of source size on photon energy and from the general nature of the X-ray surface brightness distribution that relativistic electrons, directly injected by the pulsar into the nebula, are not responsible for the X-rays. It is possible that plasma waves, originating at the pulsar, supply energy to the electrons throughout the nebula (Barnes and Scargle (1973)). Alternatively, the electrons may gain energy from plasma waves in the neighbourhood of the optical wisps (Rees and Gunn (1974)). They could then propagate throughout the rest of the nebula either by diffusion (Wilson (1972)) or by bulk motion of the plasma (Rees and Gunn (1974)).

3.2. Thermal Emission

Although the synchrotron radiation from relativistic electrons dominates the appearance of the nebula at X-ray energies, the shock wave expanding from the site of the original explosion should by now have sufficiently heated the swept up interstellar medium for the appropriate thermal X-rays (see section 2) to be detected.

Evidence for the existence of a soft source near the Crab Nebula is apparent in the 0.5 - 1 keV data of Toor et al (1976) (see Fig. 1). A significant 0.5 - 1 keV flux was observed after the nebular X-ray source had been covered by the Moon. When the detectors were pointed to a location 15° away from the Crab, a sharp drop in soft flux occurred. The position of the lunar limb at the time of the drop in flux is indicated in Figure 3. The residual soft flux can be explained by the presence of hot gas with a temperature of about 0.5 keV. The intensity at 1 keV is about 3% of the flux from the nebula corresponding to a luminosity $L_X \simeq 4.10^{36}$ ergs sec^{-1}. However the soft flux observed after immersion of the nebula could also be due to scattering of the synchrotron X-rays by interstellar dust or to the presence of a point source within 3° of the Crab.

Following the work of Toor et al, Charles and Culhane (1976a) have applied an algebraic reconstruction technique (ART) to the Copernicus X-ray telescope data from the Crab Nebula. The results of this analysis are shown in Figure 4. Two significant emitting regions are found to the north-west and south of the nebular X-ray source which respectively account for 4.6 ± 1.4% and 3.7 ± 1.2% of the Crab flux in the 0.5 - 1.5 keV band.

Application of equation 3 for a 900 year old SNR suggests a diameter of about 8 arc minutes while a similar value is obtained by scaling with age and distance from the observed diameter of Cas A. Hence the diffuse emission regions shown in Figure 4 may be parts of the heated shell of interstellar gas created by the shock wave from the original explosion.

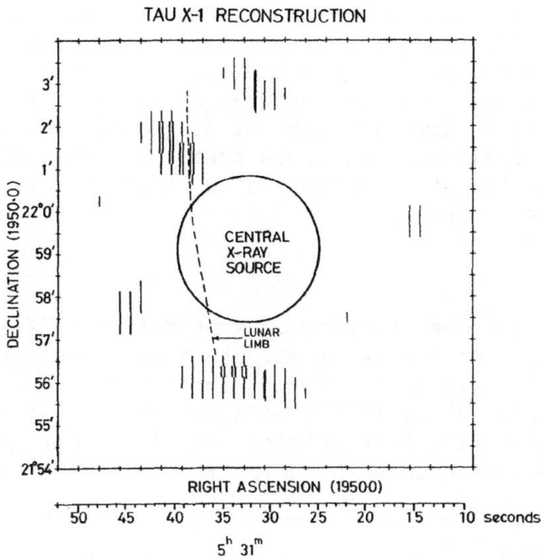

Figure 4. An 0.5-1.5 keV ART map of the region surrounding the Crab (Charles and Culhane (1976a)). The circle represents the outermost limits of the X-ray nebula. Two significant soft X-ray emitting regions are apparent.

4. YOUNG SUPERNOVA REMNANTS

We include in this category those objects which may be intermediate between phases I and II of the evolutionary scenario discussed in section 2. Excluding the Crab Nebula, the four objects in this category are listed in Table II.

Remnant Name	Age (years)	Distance (kpc)	Radius (Pc)	L_X(1-10 keV) (erg sec^{-1})	T(^oK)
Cas A	300	3	2.0	$4.5.10^{34}$ (1)	$1.3.10^7$ (1) $6.0.10^7$ (1)
Tycho	400	3	3.2	7.10^{34} (1)	$0.6.10^7$ (1) $4.1.10^7$ (1)
G287.8-0.5 (4U1043-59)	500	2.5	18	3.10^{34} (4)	$8.5.10^7$ (4)
SN 1006	970	1.2	4.4	2.10^{34} (2,3)	$1.9.10^7$ (3)

Table II Young Supernova Remnants

(1) Davison, Culhane and Mitchell (1976), Ap. J., 206, L37.
(2) Winkler and Laird (1976), Ap. J., 204, L111.
(3) Davison, Personal Communication, 1976.
(4) Becker, Boldt et al., GSFC Preprint No.X-661-76-113 (1976).

Studies of the X-ray structure of Cas A (Fabian et al (1973), Charles et al (1977) show no evidence for the presence of a pulsar at the centre of the remnant. Since the first three remnants listed in the table all exhibit emission features at 6.7 keV that are probably due to highly ionised iron (Davison et al (1976), Pravado et al (1976), Becker et al (1976)), it is probable that the X-ray emission arises from shock heated plasmas. Observations of each of thu four remnants will now be briefly reviewed.

4.1. Cas-A

This object has been studied in some detail at X-ray energies. The extended nature of the X-ray structure was first established by Fabian et al (1973) using the grazing incidence X-ray telescopes on Copernicus. Charles et al (1977) using algebraic reconstruction (ART) with the Copernicus data have obtained the 1 - 3 keV map shown in Figure 5. The emission comes principally from two bright regions that

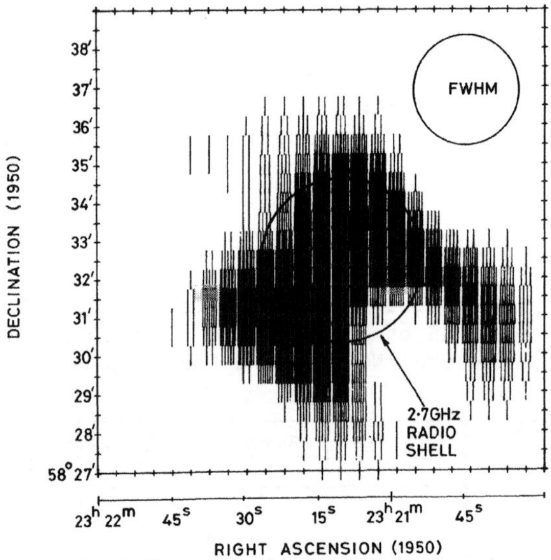

Figure 5. An ART map of Cas A (Charles et al, (1976)) showing the two X-ray emitting regions aligned on a NW - SE axis and the outline of the radio shell.

are aligned along a NW - SE axis. A less significant feature to the south-west of the shock boundary may be an artifact of variation in detector background.

Some support for the reverse shock model referred to in section 2 is provided by the X-ray spectra of the object. Recent observations suggest the presence of two plasma components at different temperatures and the existence of line emission due to highly ionised iron. (Charles et al (1975a), Pravado et al (1976a), Davison et al (1976)). The spectrum obtained by the latter workers with the Ariel V proportional counter spectrometer is shown in Figure 6a. It is of the form

$$\frac{dN}{dE} = \frac{1}{E}\left\{4.17\ \bar{g}_1\ \exp\left(-\frac{E}{1.08}\right) + 0.13\ \bar{g}_2\ \exp\left(-\frac{E}{5.15}\right)\right\}\exp\left(-\sigma(E)N_H\right) \quad (8)$$

photons cm^{-2} sec^{-1} keV^{-1}

where the $\bar{g}$ 'S are Gaunt factors and temperatures are in keV. Although no other emission features have been resolved in X-ray spectra of this object, Hill et al (1975), in fitting their data with the calculated hot plasma spectra of Tucker and Koren (1971), have concluded that emission features due to ionised Silicon must be present and that the Silicon abundance is enhanced a factor ten above its cosmic value.

4.2. Tycho

There are as yet no measurements of the X-ray structure of this object but a number of spectral studies have been made recently. A two component spectrum was first suggested by the work of Coleman et al (1973) who chose to interpret their higher energy data with a non-thermal (power law) expression. Their low energy component had a temperature of $6.3.10^6$ ^oK and suggested a need for the silicon abundance to be increased a factor 25 over the cosmic value. The spectrum of Hill et al (1975) suggests a temperature of 16.10^6 ^oK and a factor 10 enhancement of the silicon abundance. Both groups fitted their data with Tucker and Koren spectra.

The spectrum of Davison et al (1976) is shown in Figure 6b. Like Cas-A, two thermal components with temperatures of 0.55 keV and 3.5 keV are required. An emission feature due to highly ionised iron is also resolved.

4.3. G.287.8 - 0.5 (4U1043-59)

A survey of the galactic plane by Uhuru (Forman et al (1976)) lead to the detection of this source. The Uhuru error box intersects the radio contours of G.287.8 - 0.5. The source was studied by the OSO-8 proportional counter experiment (Becker et al (1976)) who found that the X-ray flux agreed with the Uhuru value thus giving support to the identification. The spectrum obtained by Becker et al (1976) is shown in Figure 6c. The high temperature of the best fitting thermal spectrum ($8.7.10^7$ ^oK) and the existence of an ionised iron emission feature argue strongly for the identification with G.287.8 - 0.5 and for the classification of this object as a young remnant.

4.4. SN 1006

This object was first identified as an X-ray source by the MIT instrument on OSO-7 (Winkler and Laird (1976)) who found the best fitting

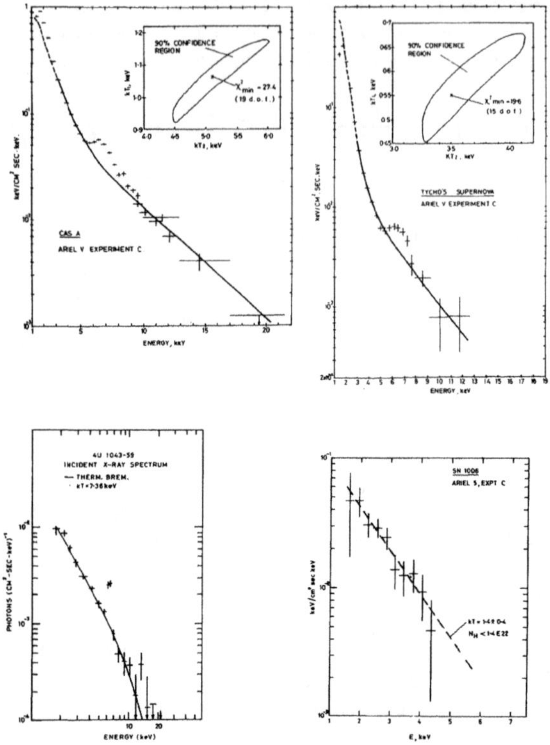

Figure 6. Proportional counter spectra of four young supernova remnants a) Cas-A, b) Tycho (Davison et al (1976), c) 4U1043-59 (Becker et al (1976), d) SN1006 (Davison (1976)).

thermal spectrum to have a temperature of about $4.5 \pm 1.9.10^7$ °K. The spectrum obtained from a recent Ariel V observation (Davison (1976)) is shown in Figure 6d. The temperature of $1.9 \pm 0.2.10^7$ °K which agrees, within the stated error, with the value obtained by Winkler and Laird, is rather low for the production of iron emission lines.

4.5. Summary

The X-ray properties of the four remnants discussed in this section are all adequately described by the ideas outlined in section 2. The bulk of the X-ray emission arises from shock heated plasma and there is as yet no evidence for the presence of pulsars or the operation of non-thermal radiation mechanisms. However it must be emphasised that the model X-ray spectra (both Tucker-Koren and simple thermal continuum) used to fit the observations, have all contained assumptions appropriate to a balanced coronal plasma. The situation must be a good deal more complex in a young supernova remnant. The ion and electron temperatures may not be equal. A coronal balance between ionisation and recombination rates may not yet have been established and the recombination times may exceed the time scales for expansion. Nevertheless continued X-ray observations of these objects with improved spatial and spectral resolution will allow us to determine whether the bulk of the shock heated material is interstellar or has been ejected in the explosion. In either case, the abundance estimates which X-ray spectroscopy will permit will be of considerable interest.

5. OLD SUPERNOVA REMNANTS

Objects in this category are in phase II of the evolutionary sequence (the adiabatic phase) and so their X-ray emission arises from shock heated interstellar plasma. The diameters obtained from radio observations are generally greater than 20 Pc and plasma temperatures, as deduced from X-ray data, are generally less than 10^7 °K. While emission lines have not yet been resolved at X-ray wavelengths, the detection of the visible coronal forbidden lines of Fe X and Fe XIV from the Cygnus Loop (Woodgate et al (1974)) and the Vela remnant (Woodgate et al (1975)) increases confidence in the applicability of the hot plasma model.

The properties of a number of old remnants are listed in Table III. For these objects, the X-ray structure and spectrum have been studied in some detail and the remainder of this section will deal mainly with the results of these observations. Several other remnants have been suggested as X-ray sources on the basis of one or two sightings. A brief summary of remnants in this category will be presented later.

Remnant Name	Age (years)*	Distance (kPc)	Diameter (Pc)	L_X(0.1 - 2.0 keV) (erg cm^{-3})	T(°K)
Cygnus Loop	20,000	0.8	40	8.10^{35}	2 - 4.10^6 °K
Vela-X	13,000	0.5 - 1.0	20 - 40	3.10^{35}	2.5 - $4.3.10^6$ °K
IC443	6,000	1.5 - 2.5	20	$0.4.10^{35}$	10.10^6 °K
Puppis-A	5,000	1.2 - 2.2	10 - 20	10.10^{35}	3 - 8.10^6 °K

Table III. Properties of Old Supernova Remnants

* An "X-ray" age estimate deduced from a plasma temperature based shock velocity (equations 4 and 5)

5.1. Cygnus Loop

Several studies have now been made of both the X-ray structure and spectrum of this object. Recent work has involved multiple scans of the source by rocket borne detectors with fields of view that are narrow in one dimension (Stevens and Garmire (1973)), Rappaport et al (1974)). Syntheses of the multiple scan one dimensional data by various techniques have been used to produce X-ray maps. An example, from the work of Rappaport et al is shown in Figure 7. Although differing in

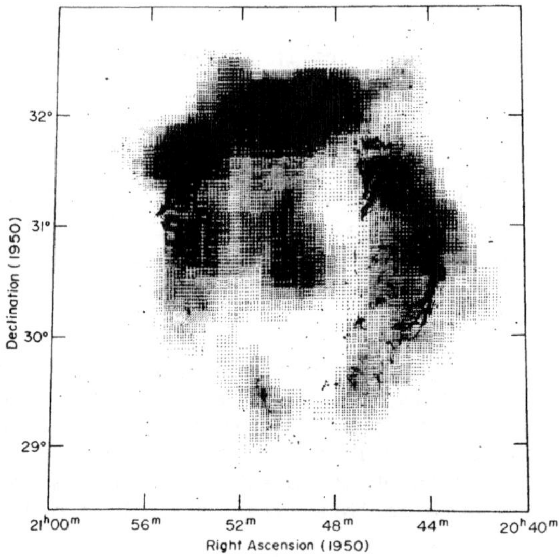

Figure 7. An ART map of the Cygnus Loop at energies below 1 keV. (Rappaport et al (1974)).

detail from the map of Stevens and Garmire, the data confirm the fragmented shell structure of this remnant. Observations by Charles et al (1975b) with the Copernicus X-ray collectors have demonstrated a correlation between optical filaments and X-ray emission in NGC6992-5 on a scale of 10 arc min but such a relationship does not exist throughout the remnant (see Figure 7). A correspondence of X-ray and radio minima at Dec. (1950) = 31° 14' in NGC6992-5 is reported by Charles et al and Gronenschild et al (1976). The latter workers used the 30' field of view X-ray collectors on the ANS spacecraft to map the source.

The X-ray spectrum of the entire Cygnus Loop has been studied by many workers. All observations favour a thermal over a power law representation of the spectrum with temperatures in the range $2\text{-}4.10^{6}$ $^{\circ}$K and values of gas column density close to 5.10^{20} H atoms/cm^2 column. A good deal of discussion has centered on the possible existence of X-ray lines in the spectrum of this object. Gorenstein et al (1971) and Stevens et al (1973) find evidence for line emission due to OVII and OVIII ions. Burginyon et al (1975) in fitting their data with the calculated hot plasma spectra of Tucker and Koren find that an increase of a factor 45 in the cosmic abundance of silicon and a decrease of a factor 5 in the iron abundance substantially improves the fit. At the required temperatures, the Tucker and Koren spectra also include significant line emission from OVII and OVIII. However all of these observations were carried out with proportional counter detectors which have considerable difficulty in resolving emission lines at energies below 2 keV. Nevertheless the detection of the coronal Fe XIV line at 5303 Å by Woodgate et al (1974) lends substantial support to the shock heated plasma models of the remnant. This observation has recently been confirmed by Woodgate and Kirshner (1976) who have in addition detected the line outside the optical filaments.

Thus it appears that the X-ray emission from the Cygnus Loop is due to shock heated interstellar plasma. Stevens (1973) has related the variation in X-ray surface brightness around the circumference of the remnant to changes in interstellar density gradients. While his model may be oversimplified (for example it does not predict the observed X-ray spectrum) it is probably qualitatively correct. However the apparent discrepancy between the filamentary expansion velocity ($\sim$120 Km sec^{-1}) and the larger shock velocity required by equation (5) to explain the plasma temperature of 2 - 4.10^6 ^{o}K ($\sim$400 Km sec^{-1}) in a uniform interstellar medium poses a problem. In particular the discrepancy leads to a considerable difference in the age of the object (18,000 years as against 50,000 years). The problem may be resolved by considering the expansion of the shock into a density gradient (Charles (1976)) and by recognising that, in the expansion of the shock through a cloudy medium, only the intercloud material is sufficiently heated to produce the X-rays (McKee and Cowie (1975)). This question will be discussed more fully in section 5.3 for the case of IC443.

5.2. Vela X

This remnant has not been as thoroughly studied as the Cygnus Loop although it appears similar in many respects. Following its identification as an X-ray source (Seward et al (1971)), it has been mapped by Gorenstein et al (1974), Moore and Garmire (1975) and by the MIT group using the SAS-3 spacecraft (Hearn (1976)). A preliminary map produced by the MIT group is shown in Figure 8. Although the data are complex, a broken shell structure can be discerned. The correlation with optical and radio features is not strong.

While the pulsar, PSR 0833-45, does not lie in the centre of the remnant, there is little doubt about the association of the two objects. The pulsar has been identified as a discrete X-ray source by Uhuru (Kellogg et al (1973), Giacconi et al (1974)) although a factor 3 difference in intensity exists between the catalogue listing and the published spectrum. Culhane et al (1974) have also detected the pulsar at lower energies. While Harnden and Gorenstein (1973) observed X-ray

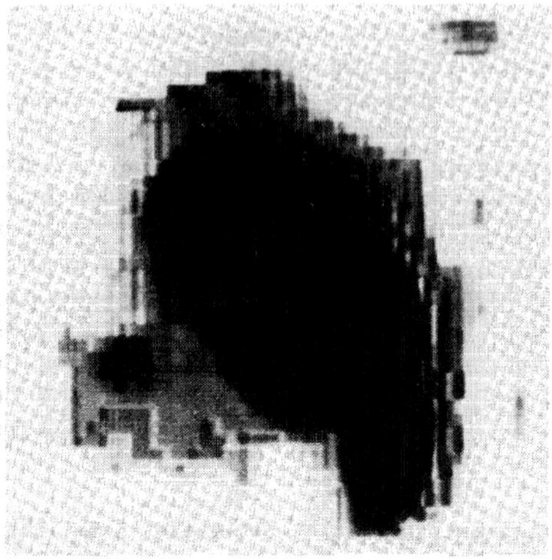

Figure 8. Soft X-ray map (0.8 - 1.2 keV) of the Vela remnant from the low energy collector on the SAS-3 spacecraft (Hearn (1976)).

pulsations in the 0.5 - 1.5 keV band, Moore et al (1974) and Rappaport et al (1974) later obtained upper limits that were substantially below the level detected by Harnden and Gorenstein. More recently, Pravado et al (1976b) have established an upper limit of 8% on the 2 - 10 keV pulsed fraction. They have also obtained a spectrum of higher intensity than that of Kellogg et al (1973).

While there are fewer observations, the situation regarding measurements of the X-ray spectrum of Vela X is similar to that described above for the Cygnus Loop. The proportional counter data strongly favour a thermal spectrum with temperatures that range between 2.6 and $4.3 \cdot 10^6$ °K (Seward et al (1971), Gorenstein et al (1974), Burginyon et al (1975), Moore and Garmire (1976)). The data have been fitted with both continuous and Tucker-Koren calculated plasma spectra. While the data of Burginyon et al are consistent with a factor 45 increase in silicon abundance, Moore and Garmire find this unacceptable but require a decrease of a factor 3 in OVII and OVIII line emission. The latter workers find two regions of the remnant to be at a significantly lower temperature than the remnant as a whole. All of the data suggest N_H values in the range $1.0 - 3.10^{20}$ H atoms/cm^2 column. Compelling evidence for the presence of hot plasma in this source is provided by the Fe XIV line observations of Woodgate et al (1975).

The broken shell structure of the hot gas is once again probably due to interstellar clouds and gas density gradients. So far no attempt has been made to incorporate the effect of density inhomogeneities into models of the shock expansion for this remnant. However the good agreement between the "X-ray" age and the age estimated from the pulsar spin down rate could indicate that the object is in evolutionary phase II with shock expansion taking place into a rather uniform interstellar medium.

5.3. IC443

Both this object and Puppis-A differ from Cygnus Loop and Vela X in being smaller, possibly younger and having higher plasma temperatures Because of its low X-ray luminosity, studies of IC443 are difficult but several recent observations have added considerably to the available information.

The remnant was first suggested to be an X-ray source in the Uhuru catalogue (Giacconi et al (1974) and the identification was later confirmed by Winkler and Clark (1974) using OSO-7. Information on the X-ray structure of the remnant has been obtained with Copernicus (Charles et al (1975c)), by a rocket observation (Malina et al (1976)) and with Ariel-V (Parkes et al (1976)). All of these observations are summarised in Figure 9 and suggest that most of the X-ray emission originates from a region just inside the prominent optical filaments to the North-East of the remnant.

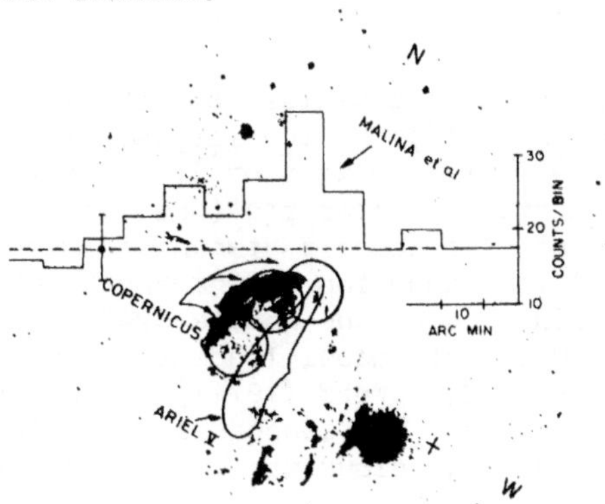

Figure 9. X-rays from IC443 showing the location of the emitting regions by Copernicus (Charles et al (1975c)), a Berkeley rocket flight (Malina et al (1976)) and Ariel-V (Parkes et al (1976)).

Malina et al (1976) have obtained a spectrum in the energy range 0.5 - 2 keV from which they deduce a temperature of 10^{+18}_{-4} 10^6 °K and a gas column of 3.5 $\pm$ 2.10^{21} atoms cm^{-2}. A comparison of the derived column density with interstellar reddening data allows a lower limit of 1 kpc to be set on the distance of the object, thus permitting the association with the HI nebula S240 at 1.5 kpc (Sharpless (1959)). The Ariel V spectrum obtained in the range 1 - 5 keV by Parkes et al allows a more precise value of temperature (10 $\pm$ 2.10^6 °K) to be estimated with the aid of the N_H value determined by Malina et al. Both spectra (see Figure 10a) are in excellent agreement in their region of over-lap. Parkes et al establish an upper limit to the value of Fe XXV emission flux which is consistent with the 10^7 °K temperature value. The existence of an excess flux at below 1 keV suggested by Shulman et al (1976) is not supported by the observation of Malina et al.

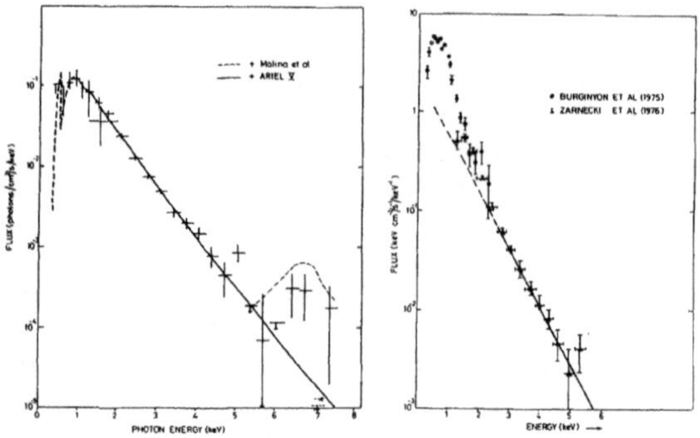

Figure 10. Spectra of two old supernova remnants a) IC443 from Malina et al (1976) and Parkes et al (1976) b) Puppis-A from Burginyon et al (1975) and Zarnecki et al (1976).

A severe discrepancy exists between the shock velocity deduced from the X-ray data ($\simeq$800 km sec^{-1}) and the largest filamentary velocity deduced by Al-Sabti (1970) ($\simeq$100 km sec^{-1}). This results in a corresponding discrepancy in the estimated age (6000 years compared to 50000 years). Parkes et al discuss a "two sector" model of the remnant in which they consider the expansion of the shock into a steep density gradient in the neighbourhood of the nebula S240 using modified Sedov equations (eqns. 7 - 10). This model leads to an age of 13,000 years for both sectors (see also Gulliford (1974)) and can account for the absence of X-ray emission from the second sector. The X-ray emission would be generated in the intercloud medium in the first sector (McKee and Cowie (1975)).

Both X-ray based estimates of age (6000 and 13000 years) would seem to exclude an association with the pulsar PSR 0611+22 whose spin down rate implies an age of 90,000 years (Graham and Hunt (1973)).

5.4. Puppis-A

While the X-ray and radio appearance of this object is not dissimilar to that of IC443, only a very few optical filaments can be seen in Pup-A. The reasons for this difference are not at all clear. Following the identification of the remnant as an X-ray source by Palmieri et al (1971), observations of its structure have been made by Zarnecki et al (1973), Gorenstein et al (1974), and Catura and Acton (1976). The Copernicus data presented by Zarnecki et al (Figure 11a) show that there is no obvious relationship between the X-ray and radio surface brightness distributions and, in particular, a small region of high X-ray surface brightness is located in the North-east of the remnant and well separated from the radio peak. In addition there is very little X-ray emission in the south and south-eastern regions. Catura and Acton have confirmed the existence of the small bright

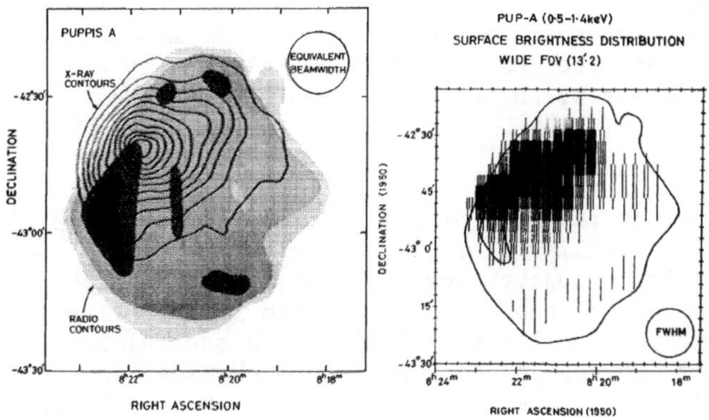

Figure 11. X-ray maps of the Puppis-A remnant in the 0.5 - 1.4 keV range a) Copernicus X-ray contour map (Zarnecki et al (1973) (no deconvolution) superimposed on the radio source (Milne (1971), b) ART map from the Copernicus data (Charles and Culhane (1976b).

X-ray region in the north-east and have, in addition, demonstrated that its spectrum is softer than that of the rest of the object. Charles and Culhane (1976b) have used ART (see sections 3 and 4) to process the Copernicus data. The resulting map, shown in Figure 11b, is in essential agreement with the uncorrected contour map in Figure 11a but shows additional structure.

Some differences exist between the observations of the Puppis-A spectrum carried out by a number of observers. With the exception of the low value of N_H obtained by Burginyon et al (1975), which is probably due to contamination from the softer spectrum of the nearby Vela X remnant, all observations of N_H are in reasonable agreement. However the temperature estimates range from 3 to 8.10^6 °K. In the energy range below 3 keV, observers using large area thin plastic windowed mechanically collimated detectors obtain values of temperature around 3 - 4.10^6 °K while observers employing X-ray focussing collectors with small fields of view obtain values around 7 - 8.10^6 °K. Observations with a large area Beryllium windowed detector on the Ariel-V satellite (Zarnecki et al (1976)), lead to a temperature of $8.6.10^6$ °K for the energy range 3 - 6 keV. This spectrum is shown in Figure 10b along with the spectrum of Burginyon et al (1975). Taken at face value, these data suggest that there are two thermal components in the Puppis-A X-ray spectrum. However the spectra obtained by Gorenstein et al (1974) and Charles et al (1975d) below 1.5 and 3 keV respectively indicate higher temperature values. In addition it is possible to explain all of the data in Figure 10b as due to the line and continuum emission from an 8.10^6 °K plasma. Further observations or analyses are urgently needed to clarify this situation.

The faint optical filaments that have been observed do not permit satisfactory estimates to be made of the expansion velocity. A straightforward application of the Sedov solution in a uniform interstellar medium leads to an age estimate of 5000 years.

5.5. Summary

All of the old remnants discussed above are well explained by the shock heated interstellar plasma hypothesis (phase II). While most discussions of these objects have previously assumed spherically symmetrical expansion of the shock into a uniform interstellar medium (e.g. Gorenstein, Harnden and Tucker (1975), Clark and Culhane (1976)), it is apparent that some of the observations described in the present review (e.g. Zarnecki et al (1973) Rappaport et al (1974), Parkes et al (1976), Moore and Garmire (1975), Charles (1976)) require that allowance be made for non-uniformities and density gradients in the interstellar medium before sensible estimates of total energy and age can be made. The difference between ages estimated from the shock temperatures and from filamentary expansion velocities can be resolved by assuming that the X-ray emission arises in the intercloud medium and by proper treatment of shock expansion into a density gradient.

6. OTHER SUPERNOVA REMNANTS

A number of well observed remnants have been discussed in the previous sections. Several other objects exist that may be X-ray sources on the basis of one or two sightings. A few of these are reasonably well established but little or no detailed data exist for the spectra and structure of any of these sources.

The names and properties of these remnants are summarised in Table IV with the exception of W66 all are apparently in phase II on the basis of their radio size. Bunner (1976) finds the temperature of the emission from the vicinity of the Lupus Loop to be about $3.5.10^6$ °K but the region is a confused one.

Source Name	Downes Catalogue Number	Galactic Co-ordinates l^{II}	b^{II}	Distance (kPc)	Diameter (Pc)	X-ray flux luminosity (erg sec^{-1})	References
MSH14-63, RCW84	36	315.4°	-2.3°	3	45	2.10^{36}	Seward et al (1975) Shulman et al (1976)
MSH15-52A, B	38, 39	320.4°	-1.2°	3	24	$1.5.10^{36}$	Seward et al (1975) Giacconi et al (1974)
Lupus Loop	45	330°	+15°	0.6	50	10^{34}	Seward et al (1975) Bunner (1976)
3C392	-	34.6°	-0.4°	1.5	14	31.10^{35}	Schwartz et al (1972) Zarnecki et al (1975)
W66	101	78.1°	+1.8°	8	3.4	10^{36}	Seward et al (1975)
HB21	108	89°	+5°	1.1	35	8.10^{34}	Seward et al (1975)

Table IV. Possible X-ray Emitting Supernova Remnants.

Although it is possible that certain large scale features in the galaxy such as the radio loops and the diffuse soft source in Eridanus (Naranan et al (1976)) may be due to old and unusually energetic supernovae, a discussion of such objects is beyond the scope of this review.

References

Al-Sabti, A.L., (1970), Ph.D. Thesis. University of Manchester.
Barnes, A. and Scargle, J.D., (1973), Ap. J., 184, 251.
Becker, R., Boldt, E., et al., (1976), GSFC Preprint No. X-661-76-113.
Bunner, A.N., (1976), Paper presented at 148th AAS Meeting, Haverford, Penn., June 21 - 24, 1976.
Burginyon, G.A., Hill, R.W., and Seward, F.D., (1975), Ap. J., 200, 163.
Catura, R.C., and Acton, L.W., (1976), Ap. J. Lett. In press.
Charles, P.A., Culhane, J.L., et al., (1975a), Ap. J., 197, L61.
Charles, P.A., Culhane, J.L., and Zarnecki, J.C., (1975b), Ap. J., 146, L19.
Charles, P.A., Culhane, J.L., and Rapley, C.G., (1975c), Astrophys. Lett., 16, 129.
Charles, P.A., Culhane, J.L., et al., (1975d), M.N.R.A.S., 170, 61P.
Charles, P.A., and Culhane, J.L., (1976a). In press. Ap. J.
Charles, P.A., and Culhane, J.L., (1976b). In preparation.
Charles, P.A., Culhane, J.L., and Fabian, A.C., (1977) In press. M.N.R.A.S.
Charles, P.A., (1976), Ph.D. Thesis. University College London.
Chevalier, R., (1975), Ap. J., 200, 698.
Clark, D.H., and Culhane, J.L., (1976), M.N.R.A.S., 175, 573.
Coleman, P.L., Bunner, A.N., et al., (1973), Ap. J., 185, L121.
Cox, D.P., (1972), Ap. J., 178, 159.
Culhane, J.L., Cruise, A.M., et al., (1974), Ap. J., 190, L9.
Davison, P.J.N., Culhane, J.L., and Morrison, L.V., (1975), Nature, 253, 610.
Davison, P.J.N., Culhane, J.L., and Mitchell, R.J., (1976), Ap. J., 206, L37.
Davison, P.J.N., (1976), Personal Communication.
Fabian, A.C., Zarnecki, J.C., and Culhane, J.L., (1973), Nature Phys. Sci., 242, 18.
Forman, W., Jones, C., and Tananbaum, H., (1976), SAO Preprint.
Fukada, Y., et al., (1975), Nature, 255, 465.
Giacconi, R., Murray, S., et al., (1974), Ap. J., Suppl. 27, 37.
Gorenstein, P., Harris, B., et al., (1971), Science, 172, 369.
Gorenstein, P., Harnden, F.R., and Tucker, W.H., (1974), Ap. J., 192, 86.
Graham, D., and Hunt, G., (1973), Nature Phys. Sci., 242, 86.
Gronenschild, E.H.B.M., Mewe, R., et al., (1976), Astron. and Astrophys., 49, 153.
Gull, S.F., (1973), M.N.R.A.S., 162, 135.
Gull, S.F., (1975), M.N.R.A.S., 171, 263.
Gulliford, P., (1974), Ap. Space Sci., 31, 241.
Harnden, F.R., and Gorenstein, P., (1973), Nature Phys. Sci., 241, 107.
Hearn, D., (1976), Private Communication.
Hill, R.W., Burginyon, G.A., and Seward, F.D., (1975), Ap. J., 200, 158.

Kafatos, M., and Tucker, W.H., (1972), Ap. J., 175, 837.
Kellogg, E., Tananbaum, H., et al., (1973), Ap. J., 183, 935.
Kestenbaum, H.L., Ku, W., et al., (1975), Ap. J., 202, L21.
Ku, W., Kestenbaum, H.L., et al., (1976), Ap. J., 204, L77.
Landau, L., and Liftshitz, L., (1959), Fluid Mechanics, Addison - Wesley Publishing Company.
McKee, C.F., (1974), Ap. J., 188, 335.
McKee, C.F., and Cowie, L.L., (1975), Ap. J., 195, 715.
Malina, R., Lampton, M., and Bowyer, C.S., (1976), Ap. J. In press.
Mewe, R., (1972), Solar Phys., 22, 439.
Milne, D.K., (1971), Aust. J. Phys., 24, 429.
Moore, W.E., Agrawal, P.C., and Garmire, G.P., (1974), Ap. J., 189, L117.
Moore, W.E., and Garmire, G.P., (1975), Ap. J., 199, 680.
Moore, W.E., and Garmire, G.P., (1976), Ap. J., 206, 247.
Naranan, S., Shulman, S., et al., (1976), NRL Preprint.
Novick, R., Berthelsdorf, R., et al., (1972), Ap. J., 174, L1.
Palmieri, T.M., Seward, F.D., et al, (1975), Ap. J., 202, 494.
Parkes, G., Charles, P.A., et al., (1976), M.N.R.A.S., In press.
Pravado, S.H., Becker, R.H., et al., (1976), Ap. J., 206, L41.
Pravado, S.H., Becker, R.H., et al., (1976), GSFC Preprint No.X-611-76-98
Rappaport, S., Doxsey, R., et al., (1974), Ap. J., 194, 329.
Rees, M.J., and Gunn, J.E., (1974), M.N.R.A.S., 167, 1.
Ricker, G.R., Sheepmaker, A., et al., (1975), Ap. J., 197, L83.
Schwartz, D.A., Bleach, R.D., et al., (1972), Ap. J., 173, L51.
Sedov, L., (1959), Similarity and Dimensional Methods in Mechanics, New York, Academic Press.
Seward, F.D., Burginyon, G,A., et al., (1971), Ap. J., 169, 515.
Seward, F.D., et al., (1975), Lawrence Livermore Laboratory Preprint.
Seward, F.D., Burginyon, G.A., et al., (1976), Ap. J., 205, 238.
Sgro, A.G., (1975), Ap. J., 197 621.
Sharpless, S., (1959), Ap. J., Suppl., 4, 257.
Shulman, S., Naranan, S., et al., (1975), Astrophys. Lett., 17, 15.
Shulman, S., Cruddace, R., et al., (1976), Paper presented to W.G.3. Session III A.2. COSPAR, Philadelphia, June 15.
Solinger, A., Rappaport, S., and Buff, J., (1975), Ap. J., 201, 381.
Staubert, R., et al., (1975), Ap. J., 201, L15.
Stevens, J.C. , (1973), Ph.D. Thesis, Caltech.
Stevens, J.C., and Garmire, G.P., (1973), Ap. J., 180, L19.
Stevens, J.C., Riegler, G., and Garmire, G.P., (1973), Ap. J., 183, 59.
Taylor, G.I., (1950), Proc. Roy. Soc., A201, 159.
Toor, A., Palmieri, T.M., and Seward, F.D., (1976), Ap. J., 207, 96.
Toor, A., and Seward, F.D., (1976), In preparation.
Tucker, W.H., and Koren, M., (1971), Ap. J., 168, 283.
Weisskopf, M., Cohen, G.G., et al., (1976), Columbia University Preprint No. 125.
Wilson, A.S., (1972), M.N.R.A.S., 160, 355.
Winkler, P.F., and Clark, G.W., (1974), Ap. J., 191, L67.
Winkler, P.F., and Laird, F.N., (1976), Ap. J., 204, L111.
Wolff, R.S., Kestenbaum, H.L., et al., (1975), Ap. J., 202, L15.
Woodgate, B.E., Stockman, H.S., et al., (1974), Ap. J., 188, L79.
Woodgate, B.E., Angel, J.R.P., et al., (1975), Ap. J., 200, 715.

Woodgate, B.E., and Kirshner, R.P., (1976), Private Communication.
Zarnecki, J.C., Culhane, J.L., et al., (1973), Nature, Phys. Sci., 243, 4.
Zarnecki, J.C., Stark, J.P., et al., (1975), M.N.R.A.S., 173, 103.
Zarnecki, J.C., Culhane, J.L., et al., (1976), In preparation.

WAS SN 1054 A TYPE II SUPERNOVA?

Roger A. Chevalier
Kitt Peak National Observatory

INTRODUCTION

The Crab Nebula is the most studied of all supernova remnants, yet it has not been possible to reconcile its properties with either of the two commonly observed types of supernovae. This contribution is an attempt to do so, taking into account the information which is available on the present nebula, the pulsar, and the original supernova event in 1054.

THE SUPERNOVA

Minkowski (1971) has claimed that the Chinese observations of SN 1054 are consistent with neither a type I nor a type II light curve. The observations of SN 1054 can be summarized as follows: it ceased to be visible in the daytime 23 days after discovery and was no longer visible at night after 653 days. This implies a decline of less than 8.5 magnitudes between these times. Thus it had a less steep decline than SN Is, which drop by about 11 magnitudes during the same period.

Minkowski dismissed the possibility of its being a SN II on the belief that these are characterized by a rapid decline. However, recent observations of SN IIs show that this is not the case. In particular, SN 1969ℓ in NGC 1058 was observed to decline slowly for close to 400 days (Ciatti et al. 1971). During the last 300 days, it declined less steeply than a SN I and an extrapolation of the light curve would be compatible with the observations of SN 1054. Other SN IIs (e.g., SN 1970 in M101) have also been observed to decline slowly at late times. Shklovsky (1968) suggested that the apparent magnitude of SN 1054 at maximum was -5. With 1.6 magnitudes of absorption (Miller 1973) and a distance of 2 kpc (Trimble 1973), this yields $M_V \approx -18$ at maximum, a value which is consistent with observations of extragalactic SN IIs (Kowal 1968).

David N. Schramm (ed.), Supernovae. 53-61.

Observations of supernovae in external galaxies show that SN IIs occur in spiral arms (Maza and van den Bergh 1976); an initial mass >4 $M_\odot$ can be deduced. Arnett (1975) finds that the He overabundance in the Crab Nebula filaments is compatible with an initial stellar core mass, M_c, in the range 2.25 to 4 $M_\odot$, corresponding to a total stellar mass of 10 to 15 $M_\odot$. Arnett claims that the central neutron star must be massive (1.45 $M_\odot$), or there would be an observable overabundance of heavy elements in the Crab filaments. Thus the mass ejected from the core may be in the range 0.8 to 2.5 $M_\odot$. Gott et al. (1970) deduced a presupernova mass of 12 $M_\odot$ on the hypothesis that it was a runaway from the I Gem association. This mass is consistent with the previous estimate and with a SN II origin for the Crab.

The hydrodynamics of the explosion of SN 1054 can be discussed with reference to the calculations of Chevalier (1976). The core collapse resulted in the deposition of about 2 x 10^{51} ergs at the center of the He-rich outer parts of the core. The core material formed a cool shell, which collided with the envelope. Since the envelope mass was $>M_c$, the core mass was decelerated by the envelope and a high pressure region was built up. This resulted in a reverse shock wave which drove the core material back to the center. The original deceleration of the core shell was Rayleigh-Taylor unstable, so some mixing between envelope and core material is expected at that time. If only 0.5 $M_\odot$ of envelope material was mixed with the core, the abundance observations can be approximately reproduced (Arnett 1975). The implosion of the material mixed it thoroughly. The supernova energy was transferred to the envelope which expanded with a mean velocity of 5000 km/s. The envelope dominated the light in the tens of days around maximum light. It then became optically thin and the now slowly expanding core material contributed the observed light. Numerical calculations have shown that the light curve can continue for several hundred days in a way consistent with observations of SN IIs (Chevalier 1976).

In this picture, the central pulsar plays a small role during the supernova explosion. The initial energy input of the pulsar, L_i, was about 10^{40} erg/s (see next section). Assuming an effective temperature of 10,000° at maximum light, the supernova luminosity was about 10^{43} ergs/s. It is unlikely that the pulsar contributed a significant part of the observed luminosity near maximum light. Yet it did have an important dynamical effect in the following hundreds of years.

THE INNER NEBULA

In this section, we explore the hypothesis that what is now called the Crab Nebula resulted from the interaction of the central pulsar with the slow moving core material described in the previous section. The observational parameters to be considered are the present pulsar rotation rate Ω, its time derivative $\dot{\Omega}$, the braking index $n(\equiv \Omega\ddot{\Omega}/\dot{\Omega}^2)$, the rms velocity of the filaments v_{rms}, the age of the nebula, and the convergence time of the filaments. We have $\Omega = 190\ s^{-1}$, $\dot{\Omega} = -2.42 \times 10^{-9}\ s^{-1}$ (cf. Taylor and Manchester 1975), n = 2.5 (Groth 1975), $v_{rms} \approx 1400\ km\ s^{-1}$ (Woltjer 1972), and Δt (age-convergence time) = 86 ± 10 yrs (Trimble 1968).

As a result of the supernova explosion, the rms velocity of the core material is about an order of magnitude smaller than the present v_{rms}. The present kinetic energy of the Crab must have been derived almost entirely from the pulsar energy. As noted above, the nebular mass may be in the range 1.3 to 3.0 $M_\odot$ (ejected core mass plus 0.5 $M_\odot$ of H-rich material). It has not been possible to place useful limits on the mass from observations; Davidson et al. (1974) note that any mass in the range 0.3 to 10 $M_\odot$ is allowed. The present kinetic energy of the remnant is about $4 \times 10^{49}\ M_n$ ergs, where M_n is the nebular mass in units of 2 $M_\odot$. The energy given up by the pulsar to the present time, E_ℓ, must be greater than this.

If n is constant, Ω and L, the pulsar power output, evolve as follows:

$$\Omega = \frac{\Omega_i}{(1+t/\tau)^{1/(n-1)}} \quad , \quad L = \frac{L_i}{(1+t/\tau)^{(n+1)/(n-1)}} \tag{1}$$

and the initial pulsar energy can be found from

$$E_i = \frac{1}{2} I\Omega^2 \left[1 - \frac{t(n-1)\dot{\Omega}}{\Omega}\right]^{-2/(n-1)} \tag{2}$$

where Ω and $\dot{\Omega}$ refer to the time t and I is the neutron star moment of inertia. The assumption of constant n is probably not valid; n may have been higher in the past. Therefore, in the Table, the pulsar parameters are calculated for several values of n. The range is theoretically plausible. I_{45} is the moment of inertia in units of 10^{45} gm cm^2. Borner and Cohen (1973) find I_{45} = 1.9 for a 1.4 $M_\odot$ neutron star. If this is correct, the E's and L in the Table must be multiplied by 1.9. It should be noted that smaller values of the moment of inertia have been derived; for example, Baym, Pethick, and

Crab Pulsar Model Parameters

n	E_i/I_{45} (ergs)	E_ℓ/I_{45} (ergs)	L_i/I_{45} (erg s^{-1})	τ (yr)	Δt (yr)	$V_{pr}(M_n/I_{45})^{1/2}$ (km s^{-1})
2.5	5.25×10^{49}	3.45×10^{49}	2.99×10^{39}	744	215	611
3.0	6.82×10^{49}	5.02×10^{49}	6.57×10^{39}	329	186	1410
3.5	1.36×10^{50}	1.18×10^{50}	4.30×10^{40}	80	122	2420
3.72	∞		∞	0		

Sutherland (1971) find $I_{45} = 0.6$. In any case, it is quite feasible that the pulsar has provided the total kinetic energy of the nebula.

We now investigate the interaction of the relativistic bubble produced by the pulsar with the expanding stellar remnant. It is assumed that all of the pulsar power goes into a $\gamma = 4/3$ fluid for which radiative losses are unimportant. For the time being, a spherically symmetric flow is assumed. A simple model for the expanding remnant will be taken; in view of the paucity of data on the supernova, a more detailed model does not seem warranted. On the basis of hydrodynamic calculations, three regions are expected to result from the supernova explosion (see figure 1). Region 1 contains the slow-moving core mass which has been decelerated by its interaction with the envelope (this requires $M_3 \gtrsim M_1$). M_1 is the core mass not in a compact remnant. Region 3 is the high velocity shell formed from the envelope of the presupernova. It carries away most of the supernova energy. In between is a low density region with mass $M_2 << M_1, M_2$. In the simple model it is assumed that each region has a constant density. On the basis of type II supernova models and observations, the velocities at the boundaries of the regions are approximately $v_1 \approx 300$ km s^{-1}, $v_2 \approx 3000$ km s^{-1}, and $v_3 \approx 7000$ km s^{-1}. The density in Region 1 is thus 10^4 times greater than that in Region 3. It is assumed that the expansion of the material is uniform; that is, at any one time velocity is proportional to radius. The density of a gas element drops proportional to t^{-3}.

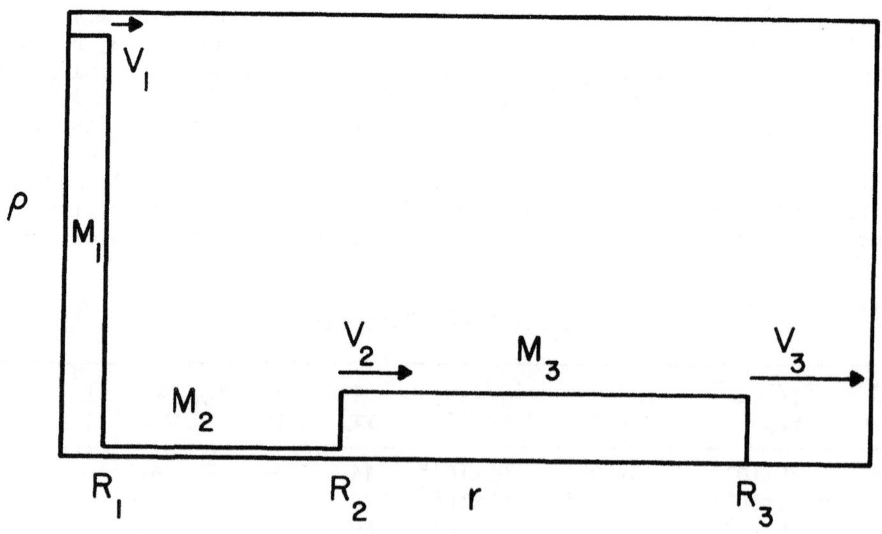

Figure 1. Simple model for post-supernova density distribution based on hydrodynamics of the explosion (not to scale).

The first stage of evolution involves the interaction of the pulsar bubble with the mass M_1. It is assumed that the pulsar L is constant during this phase, and that the gas cools after being shocked. The mass is swept up into a thin shell which can be characterized by a radius R. The equation of motion of the shell, the adiabatic expansion of the pulsar bubble, and the accretion of mass onto the shell are described by (cf. Ostriker and Gunn 1971)

$$M_s \frac{d^2R}{dt} = 4\pi R\ [P_c - \rho_1\ (\nu - v)^2] \tag{3}$$

$$\frac{d(4\pi R^4 P_c)}{dt} = {}^{\prime}R \tag{4}$$

$$\frac{dM_s}{dt} = 4\ R^2\rho_1\ (\nu - v) \tag{5}$$

where M_s is the shell mass, P_c is the pressure in the bubble, $\nu = dR/dt$, and v is the initial uniform expansion velocity, ρ_1 decreases with time as t^{-3}. The solution to these equations is

$$R = \left(\frac{125}{99}\ \frac{v_1^3\ L}{M_1}\right)^{1/5} t^{6/5}\ . \tag{6}$$

The end of the first phase occurs when $R = v_1 t$ or at

$$t_1 = \frac{99}{125}\ \frac{M_1 v_1^2}{L} = 9.3\ \left(\frac{M_1}{2M_\odot}\right)\ \left(\frac{v_1}{300\ \text{km s}^{-1}}\right)^2\ \left(\frac{L}{10^{40}\text{erg s}^{-1}}\right)^{-1}\ \text{yr.} \tag{7}$$

It does not take much time to sweep up the core mass if its initial velocity is small. During the early part of this phase, the mass M_1 is still optically thick at optical wavelengths and it is likely that the pulsar L contributes to the late time supernova light curve. Bodenheimer and Ostriker (1974) have explored pulsar-driven supernovae. The L_i taken here is much smaller than in their work.

After time t_1, the shell expands into the low density Region 2. Because $M_2 << M_1$, M_2 does not affect the dynamics of the expanding shell. The same equations listed above apply but with $\rho_1 = 0$ and $M_s = M_1$. The solution for constant L is (cf. Ostriker and Gunn 1971)

$$R = \left(\frac{8}{15}\ \frac{L}{M_1}\right)^{1/2}\ t^{3/2}\ . \tag{8}$$

We now investigate the structure of the shell on the assumption that the whole shell has an equilibrium temperature T. To begin with, assume that the pressure in the shell is equal to that in the pulsar bubble, P_c. The pressure can be obtained from the internal energy of the bubble which is 2/5 Lt. Then the thickness is

$$\frac{\Delta R}{R} = \frac{5}{2}\frac{kT}{\mu}\frac{M_1}{Lt} =$$

$$2.2\mathrm{x}10^{-3}\ (\frac{T}{10^4 K})\ (\frac{M_1}{2M_\odot})\ (\frac{L}{10^{40}\mathrm{erg\ s^{-1}}})^{-1}\ (\frac{\mu}{2\mathrm{x}10^{-24}})^{-1}\ (\frac{t}{10\ \mathrm{yr}})^{-1} \quad (9)$$

where μ is the mean molecular weight in the shell. Actually the pressure is not constant throughout the shell. It is the same as P_c on the inner boundary and is the post-shock pressure, $P_{sh} = 3/4\ \rho_2 v^2$, on the outer boundary. The pressure scale height can be calculated from the acceleration of the shell; it is the same as ΔR. Thus the actual shell thickness is $\Delta R\ \ln P_c/P_{sh}$. It is clear that with radiative cooling behind the shock wave, the shell thickness is small. It can be shown that even in the adiabatic case, it is less than 0.1 of its radius.

The next phase of evolution occurs when $R \approx v_2 t$. The observed velocities in the Crab indicate that interaction with the fast-moving shell may just be beginning. The largest diameter of the Crab Nebula is 420" arc (van den Bergh 1970), compatible with a constant velocity of 2200 km s^{-1} since the time of the explosion. The observation that the densest filaments are well inside the outer boundary of the amorphous region of radiation suggests that deceleration of the core mass by ram pressure is negligible and that the coupling of the relativistic fluid to the core mass is poor at present. Presumably the filaments formed as a result of the R-T (Rayleigh-Taylor) instability at the interface between the relativistic fluid and the accelerated thermal gas. During the $R \propto t^{3/2}$ phase of the evolution, the acceleration is $a = (3/4)\ (R/t^2)$. The development of the instability is important for scales which are similar to the thickness of the shell, ΔR. The non-linear regime will be reached if $\tau_{R-T} << t$, where τ_{R-T} is the R-T instability growth time. We have

$$\frac{\tau_{R-T}}{t} = \frac{1}{t}\ (\frac{\Delta R}{a})^{\frac{1}{2}} = (\frac{4}{3}\ \frac{\Delta R}{R})^{\frac{1}{2}} \quad . \quad (10)$$

Although the instability generally does have time to grow, equation (9) shows that it can develop most fully at later times.

We now check the compatibility of the acceleration with the observed value of Δt (Crab age - convergence time). The longest part of the evolution is described by the relativistic bubble accelerating a thermal shell into a medium of negligible mass. This is described by equations (3) and (4) with L given by equation (1). For each value of n a value of Δt can be found by integrating the equations (see the Table). Also shown in the Table are the predicted present shell velocities, v_{pr}, found from the integrations. It is evident that Δt is generally longer than the observed value. But these models do not take into account the R-T instability, which reduces the coupling between the relativistic fluid and the shell. If the model with n = 3.0

is taken with $I_{45}/M_n = 2$, and it is assumed that the acceleration stops when $v = 1400$ km s^{-1} is reached, a value of $\Delta t = 62$ years is obtained. Models with reasonable values of n and I_{45}/M_n can reproduce the observed value of Δt. The other possibility for the small value of Δt is that the initial energy of the shell was significant compared to the present energy. Such a model is discussed by Ostriker and Gunn (1971). However, the supernova luminosity may be more difficult to explain with this model.

In summary, the interaction of the pulsar with the core material proceeds as follows. The pulsar power feeds a relativistic bubble which first sweeps up the core material into a cool shell. The expanding bubble accelerates the shell until the Rayleigh-Taylor instability breaks up the shell and decreases the coupling between the bubble and the shell. The inner nebula, which is identified with the Crab, ends up with several percent of the total supernova energy. It is the higher velocity shell which carries away the dominant energy.

THE OUTER NEBULA

As is well-known, an outer nebula has not been observed. Possible observations of a 10 $M_\odot$ shell moving at 5000 km s^{-1} must be examined. The predicted column density through the shell is about 3×10^{18} atom cm^{-2}. The mean density of the shell is about 1 atom cm^{-3}. Considering that the Crab is about 200 pc from the galactic plane, the ambient density is likely to be considerably smaller. The shell is probably freely expanding and has swept up very little interstellar medium gas.

This conjecture immediately implies that the shell will be difficult to observe at radio wavelengths. Nonthermal radio emission presumably arises as a result of particle acceleration by turbulence when the supernova shell has swept up about its own mass in the surrounding medium (Gull 1973, 1975; Scott and Chevalier 1975).

Thermal X-ray emission has possibly been observed from a region surrounding the Crab Nebula (Toor et al. 1976). In the present model, there are two possible sources of thermal X-rays. First, the relativistic pulsar bubble may be interacting with the inner part of the rapidly expanding outer shell. The situation would be similar to that described by Max et al. (1976). Second, a reverse shock wave may be driven into the shell as the interaction with the surrounding medium becomes significant (McKee 1974). As the observed X-ray temperature is low (0.5 keV), the ratio of swept up mass to shell mass must be small. Neither of these models in their simple form seem capable of explaining the high observed thermal luminosity. Further observations of this emission will be of great interest.

The amount of mass ionized by the Crab Nebula can be estimated by Stromgren sphere arguments (cf. Spitzer 1968). An ionizing luminosity of 3×10^{35} erg s^{-1} yields an ionized mass of about $24/n_H$ $M_\odot$, where

n_H is the hydrogen atom number density. If the shell is fairly smooth, complete ionization is expected. The expansion of the shell should decrease the electron column density N_e: $dN_e/dt \approx 2N_e/t_o \approx 2\times10^8$ cm^{-2} s^{-1}, where t_o is the age of the remnant. Observationally, changes of about 10^{10} cm^{-2} s^{-1} can be measured and much larger fluctuations have been observed (Rankin and Roberts 1971), so that the predicted effect is not observable.

Assuming that the gas is ionized at $T \approx 10^4$°K, the emission measure is about 5 pc cm^{-6}. This can be compared with the diffuse emission from the galactic plane, which has been measured at 4 pc cm^{-6} (Reynolds et al. 1974). The shell surrounding the Crab Nebula should be distinctive because of its spatial structure, but primarily because of its broad velocity range (5000 - 10000 km s^{-1}). The emission is thus contaminated with a considerable amount of night sky emission, but it may be detectable with modern techniques, particularly if it is clumpy. These observations will provide the best check on the present model.

DISCUSSION

If the Crab Nebula was a typical SN II, other objects should have a similar morphology. In fact there are other nonthermal radio sources with the flat spectrum and the amorphous structure characteristic of the Carb. The best studied of these is 3C 58 (Wilson and Weiler 1976), which is perhaps the result of SN 1181 (Stephenson 1971). The diameter of 3C 58 is five times larger than that of the Crab, implying a larger expansion velocity. If the Crab Nebula were to expand to this larger volume, its surface brightness would be considerably less than that observed in 3C 58. This would imply that the central pulsar had a larger initial energy, which is consistent with the higher velocity of the possible shell in 3C 58.

At present, the dense Crab Nebula filaments are embedded within the light relativistic fluid, so that any deceleration is negligible. As the pulsar energy input decreases and the interaction of the relativistic bubble with the surrounding medium becomes dominant, the filaments will pass out of the relativistic fluid and into the thermal gas which has been shocked by the outer shell. As the filaments move through the gas, they may be broken up by instabilities and evaporated by heat conduction. The evolution of the relativistic particle and magnetic field energies in the bubble may be described by a model like that of Rees and Gunn (1974). The expansion of the bubble will lower the surface brightness of the bubble and the energy of the outer shell will give rise to the dominant radiative effects. In its advanced stages, the evolution will be similar to models of supernova remnants which neglect the presence of a pulsar.

In conclusion, it is possible to take the conservative view that the SN 1054 was not an exceptional event, but was a normal SN II. The crucial prediction of this view is that there is an as yet undetected high velocity shell surrounding the Crab Nebula. While this suggestion has been discussed for several years (e.g., by M. Rees), it appears that the needed observations have not yet been made.

The author acknowledges helpful discussions with M. Rees and J. Arons and the hospitality of the Institute of Astronomy, Cambridge, where this paper was prepared.

REFERENCES

Arnett, W. D.: 1975, *Astrophys. J.* 195, 727.

Baym, G., Pethick, C., and Sutherland, P.: 1971, *Astrophys. J.* 170, 299.

Bodenheimer, P. and Ostriker, J. P.: 1974, *Astrophys. J.* 191, 465.

Borner, G. and Cohen, J. M.: 1973, *Astrophys. J.* 185, 959.

Chevalier, R. A.: 1976, *Astrophys. J.* 207, 872.

Ciatti, F., Rosino, L., and Bertola, F.: 1971, *Mem. Soc. Astron. Italia* 42, 163.

Davidson, K., Crane, P., and Chincarini, G.: 1974, *Astron. J.* 79, 791.

Gott, J. R., Gunn, J. E., and Ostriker, J. P.: 1970, *Astrophys. J. Letters* 160, L91.

Groth, E. J.: 1975, *Astrophys. J. Suppl.* 29, 453.

Gull, S. F.: 1973, *Monthly Notices Roy. Astron. Soc.* 161, 47.

Gull, S. F.: 1975, *Monthly Notices Roy. Astron. Soc.* 171, 263.

Kowal, C.: 1968, *Astron. J.* 73, 1021.

Max, C., Arons, J., and Blandford, R.: 1976, paper presented at the Berkeley meeting of the ASP.

Maza, J. and van den Bergh, S.: 1976, *Astrophys. J.* 204, 519.

McKee, C. F.: 1974, *Astrophys. J.* 188, 335.

Miller, J. S.: 1973, *Astrophys. J. Letters* 180, L83.

Minkowski, R.: 1971, *IAU Symposium No. 46*, eds. R. D. Davies and F. G. Smith, (Reidel, Dordrecht), p. 241.

Ostriker, J. P. and Gunn, J. E.: 1971, *Astrophys. J. Letters* 164, L95.

Rankin, J. M. and Roberts, J. A.: 1971, *IAU Symposium No. 46*, eds. R. D. Davies and F. G. Smith, (Reidel, Dordrecht), p. 114.

Rees, M. J. and Gunn, J. E.: 1974, *Monthly Notices Roy. Astron. Soc.* 167, 1.

Reynolds, R. J., Roesler, F. L., and Scherb, F.: 1974, *Astrophys. J. Letters* 192, L53.

Scott, J. S. and Chevalier, R. A.: 1975, *Astrophys. J. Letters* 197, L5.

Shklovsky, I. S.: 1968, *Supernovae*, Interscience, New York.

Spitzer, L.: 1968, *Diffuse Matter in Space*, Interscience, New York.

Stephenson, F. R.: 1971, *Quart. J. Roy. Astron. Soc.* 12, 10.

Taylor, J. H. and Manchester, R. N.: 1975, *Astron. J.* 80, 794.

Toor, A., Palmieri, T. M., and Seward, F. D.: 1976, *Astrophys. J.* 207, 96.

Trimble, V.: 1968, *Astron. J.* 73, 535.

Trimble, V.: 1973, *Publ. Astron. Soc. Pacific* 85, 579.

van den Bergh, S.: 1970, *Astrophys. J. Letters* 160, L27.

Wilson, A. S. and Weiler, K. W.: 1976, *Astron. Astrophys.* 49, 357.

Woltjer, L.: 1972, *Ann. Rev. Astron. Astrophys.* 10, 129.

THE EVOLUTION OF SUPERNOVA REMNANTS

John R. Dickel, University of Illinois Observatory
D. C. Wells, Kitt Peak National Observatory
T. R. Gull, Lockheed Elec. Corporation
A. G. Willis, Leiden Sterrewacht
and
S. van den Bergh, David Dunlap Observatory

Recent advances in image tube photography and other optical techniques together with the opportunity to obtain high resolution radio maps, have now given us the ability to follow in detail the evolution in morphology of supernova remnants. To keep our summary of this evolution brief we shall limit our discussion to only two representative examples: the very young remnant of Tycho's supernova and the much older remnant CTB1.

Let us first look at Tycho's SNR. Figure 1 is an overlay of a radio and optical picture as displayed on the Interactive Picture Processing System of Kitt Peak (Wells, 1975). The continuum radio map of this object was made at a wavelength of 11 cm with the NRAO interometer. The effective HPBW of the telescope was about 1/50 the diameter of the remnant so most of the emission is definitely amorphous arising in an irregular shell without much fine structure. By contrast the red photograph (1-meter telescope and 3 stage Carnegie image tube) reveals only a few thin filaments around the periphery of the remnant which all lie just on the outside edge of significant features in the radio shell with noticeable discontinuities. Even the filament on the northeastern side which is inside the boundary of the radio emission does lie along a brighter ridge of radio emission; it is most likely a feature toward the fore or back side of the remnant seen in projection.

This picture would suggest that the relativistic particles responsible for the synchrotron emission from this young remnant are contained in the expanding piston moving outward from the explosion center. The magnetic field in this and other young remnants has been known for several years now to have a predominantly radial structure. The optical features would represent the region of compression and heating of the surrounding interstellar medium by the shock preceding the expanding shell. The heating is sufficient to produce the observed

David N. Schramm (ed.), Supernovae. 63-65.

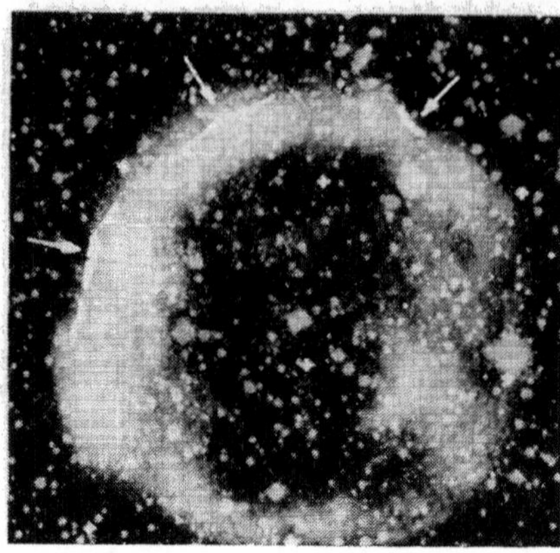

Figure 1. Superposed ratio and optical pictures of Tycho's SNR. The general amorphous shell depicts the radio emission and optical filaments lie as indicated by the arrows just along the left edge, to the upper right, and just inside the shell on the upper left.

hydrogen recombination emission but the region is still much cooler than the interior zone which produces the synchrotron and x-ray emission. The picture would suggest that the brightest optical features appear where the expansion has been the least, presumably because of density irregularities in the surrounding medium so the brightness of the filaments is most likely a reflection of the density enhancement.

In contrast to the amorphous structure in the young SNR, Tycho, let us now look at Figure 2, a radio map of CTB1 made at a wavelength of 50 cm with the WSRT. The bright spots represent probably unrelated point sources such as radio galaxies, quasars, etc. which accidentally lie in the field of the remnant. (They are, so to speak, the radio equivalent to contaminating stars). The field of view is a little over 1 degree and the effective HPBW of the synthesized beam is about 1 arcmin; there are about 6 display pixels per observed resolution element. Clearly several features remain unresolved even with this resolution and must be narrower than about 1/60 the 35-arcmin diameter of the remnant. These thin sinuous structures certainly dominate the emission and the presence of some toward the center of the remnant indicated that they are filaments rather than sheets. All of the brightest radio features have a one-to-one correspondence with optical counterparts on a 48-inch Schmidt plate (van den Bergh, et al. 1973).

The filaments themselves are not completely uniform but we get the impression of condensations at points along a magnetic field line. In general, the magnetic fields in old SNR have evolved to be sort of mixed and messy but with some tendency to align tangentially to the structural outline of the remnants (e.g. Dickel and Milne 1976).

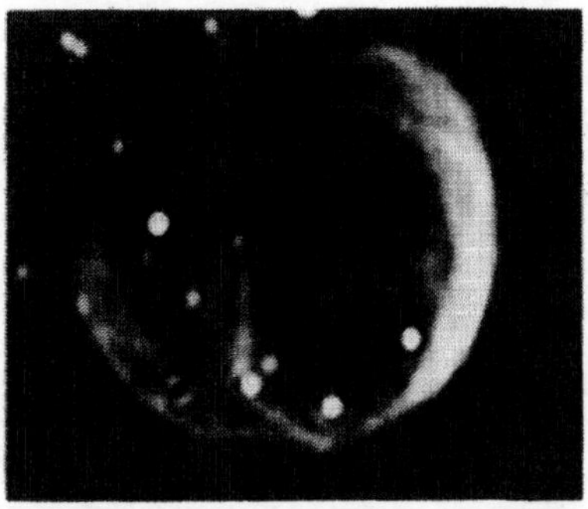

Figure 2. Radio picture of CTB1 showing its filamentary structure.

Such a picture would support the model of Duin and van der Laan (1975) in which small condensations in the swept up interstellar medium have been compressed and cooled under pressure equilibrium. The magnetic field follows the gas (the gas pressure is originally considerably stronger than the magnetic) and inhibition of conduction across field lines tend to enhance the condensation process up to factors of 10 to 100 which is sufficient to produce the observed emissivity and structures.

This presentation has been a quick glimpse of the major structural changes which occur during the evolution from a young to an old SNR.

REFERENCES

Bergh, S. van den, Marscher, A. P. and Terzian, Y. 1973, Ap. J. Suppl., 26, 19.

Dickel, J. R. and Milne, D. K. 1976, Austr. J. Phys., in press.

Duin, R. M. and van der Laan, H. 1975, Astron. Astrophys., 40, 111.

Wells, D. C. 1975, in Proceedings of the Conference of Automation in Astronomy, Cambridge, Mass.

EVIDENCE FOR A CLASS OF SUPERNOVA REMNANTS RESEMBLING THE CRAB NEBULA

K.W. Weiler and A.S. Wilson
Max-Planck-Institut für Radioastronomie, Bonn, Germany; and Astronomy Centre, University of Sussex, Falmer, Brighton, England

Supernova remnants are usually identified by their general properties of a) being Galactic, b) exhibiting relatively steep, non-thermal radio spectra, c) showing at least partial shell structure at radio wavelengths and d) sometimes being associated with expanding optical filaments. These features are found in both young (e.g. the remnant of Tycho's supernova, 3C 10, and Cassiopeia A) and old (e.g. the Cygnus loop and IC 443) remnants. Roughly 150 radio objects have been identified as members of this class.

There is now evidence for the existence of another, much smaller group of supernova remnants which resemble, at least in their radio properties, the Crab Nebula — an object previously considered unique. The general signatures of these objects are of a) being Galactic and radiating non-thermal radio emission but b) having relatively flat radio spectra and c) exhibiting "filled-centre" rather than "shell-like" radio structures. In addition, these objects show strong linear polarization and rather uniform magnetic field organization in their central regions. Finally, there is some indication that they may deviate from the well-known Surface Brightness - Linear Diameter (Σ - D) relation (which describes the evolution of the "shell-like" remnants) in the sense of being too small or greatly under luminous.

The three known members of this class are the Crab Nebula, 3C 58, and G21.5-0.9. For their radio brightness and linear polarization distributions see (e.g.) Weiler and Seielstad (1972) and Wilson (1972) for the Crab Nebula, Weiler and Seielstad (1971) and Wilson and Weiler (1976a) for 3C 58, and Becker and Kundu (1976) and Wilson and Weiler (1976b) for G21.5-0.9. In these references, the similar "filled-centre" total intensity shapes and the well organized linear polarization distributions (and thus magnetic fields) in the central regions of the three sources are clearly apparent. The relative flatness of their radio spectra (Crab, $\alpha = -0.26$; 3C 58, $\alpha = -0.09$; G21.5-0.9, $\alpha = +0.01$; $S \propto \nu^{+\alpha}$) is apparent when compared to the two young shell-type remnants 3C 10 and Cas A which have spectral indices of $\alpha = -0.55$ and $\alpha = -0.79$, respectively.

David N. Schramm (ed.), Supernovae. 67-69.

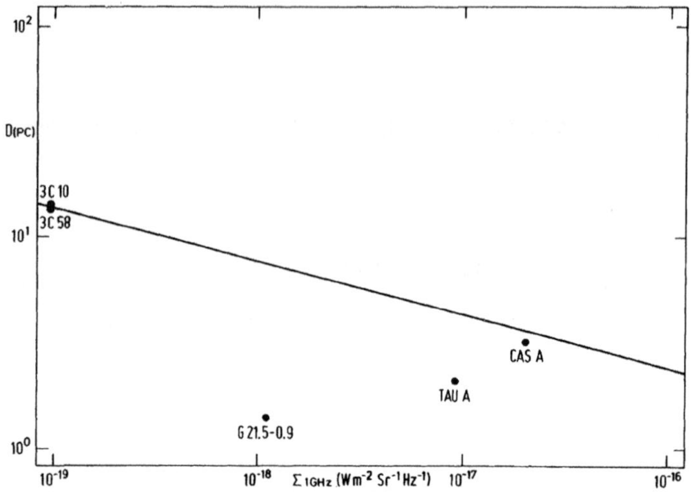

Figure 1. Illustration of the places of the three "Crab-like" supernova remnants and the two young shell remnants, 3C 10 and Cassiopeia A, on the Surface Brightness - Linear Diameter (Σ - D) diagram. The line shown is taken from Ilovaisky and Lequeux (1972).

As an illustration of their places on the Σ - D diagram, the three "Crab-like" supernova remnants are shown along with 3C 10 and Cas A in Figure 1. The line has been taken from Ilovaisky and Lequeux (1972). It may be seen that whereas 3C 10 and Cas A lie close to the line, the Crab Nebula and G21.5-0.9 lie considerably below it. 3C 58, on the other hand, at an assumed distance of 8 Kpc (see Goss et al., 1973 and Williams, 1973), is consistent with the line. Hence it is not clear whether deviation from the Σ - D relation is a common property of "Crab-like" supernova remnants.

Although neither 3C 58 nor G21.5-0.9 have been detected at other than radio wavelengths, the lack of detection of optical emission can be ascribed, at least in part, to interstellar obscuration. However, their properties could be explored further through:

1) optical studies of the stars near the centre of 3C 58 to look for a stellar remnant of the supernova explosion;

2) high sensitivity x-ray and infrared observations of both sources;

3) observations at short (λ < 2 cm) and long (λ > 75 cm) radio wavelengths to further explore their radio spectra; and

4) searches for radio pulsars having a short period and a high dispersion measure.

REFERENCES

Becker, R.H. and Kundu, M.R.: 1976, Astrophys. J. 204, 427.

Goss, W.M., Schwarz, U.J., and Wesselius, P.R.: 1973, Astron. & Astrophys. 28, 305.

Ilovaisky, S.A. and Lequeux, J.: 1972, Astron. & Astrophys. 18, 169.

Weiler, K.W. and Seielstad, G.A.: 1971, Astrophys. J. 163, 455.

Weiler, K.W. and Seielstad, G.A.: 1972, Astron. & Astrophys. 21, 393.

Williams, D.R.W.: 1973, Astron. & Astrophys. 28, 309.

Wilson, A.S.: 1972, Monthly Notices Roy. Astron. Soc. 157, 229.

Wilson, A.S. and Weiler, K.W.: 1976a, Astron. & Astrophys. 49, 357.

Wilson, A.S. and Weiler, K.W.: 1976b, Astron. & Astrophys. (in press)

IONIZATION STRUCTURE OF THE CYGNUS LOOP

Theodore R. Gull, Lockheed Electronics Company, Inc.
Robert A. R. Parker, NASA-Johnson Space Center
Robert P. Kirshner, Kitt Peak National Observatory

A survey instrument capable of narrow band filter imagery has been put into operation at KPNO. It consists of large diameter filters with 27 to 90 Angstroms (full width at half transmission), a commercially available 300 mm f/2.8 lens and a two-stage magnetically focused image intensifier. Frames having thirty arcsecond angular resolution over a nominal seven degree field of view are recorded to sky background limit in ten to fifteen minutes. This camera is being used to survey the galactic plane in Hβ,[SII] (6725 A), [OIII] (5007 A), Hα + [NII] (6570 A) and continuum at 4770 A. A preliminary program surveying special objects such as supernova remnants is near completion.

Calibrated frames have been recorded of the Cygnus Loop through the filters listed above. The Interactive Picture Processing System (IPPS) at KPNO was used to obtain star-free frames of Hβ, [OIII], Hα + [NII] and [SII]. Color composites of these frames were made. Ratios of Hβ/[OIII], (Hα + [NII])/[SII] were done.

The reduced data frames demonstrate the following:

1) A faint outer arc exists to the North and Northeast of the bright northern boundary filaments. It is seen in Hβ, and Hα + (NII).
2) Color composites of (Hα + [NII]) and [OIII] show that [OIII] is fainter on the inside of the loop.
3) Color composites of (Hα + [NII]) and [SII] show that [SII] is stronger on the inside of the loop.

David N. Schramm (ed.), Supernovae. 71.

SUPERNOVAE AND QUASARS

Stirling A. Colgate and Albert G. Petschek
New Mexico Institute of Mining and Technology
Socorro, New Mexico 87801

ABSTRACT

We attempt to show how the evolution of galactic nucleus of the special mass 10^8 $M_\odot$'s leads preferentially to the radiation phenomena of quasars. Radiation flow greater than the Eddington limit leads to the correct mass flow and lifetime. The evolution of the stellar system leads to collision and massive stars; rapid supernovae give canonical neutron stars which in turn make much larger supernovae - brand X supernovae - when a small cold neutron star coalesces with a massive star. The high energy of ejected matter, 10 Mev per nucleon, and 4 x 10^{53} ergs controls the matter density of the nebula of gas to thickness of $\simeq$ 3 Compton mean free paths and electron temperatures of $\simeq 1/2\ m_e c^2$ These conditions are ideal for plasma wave photon emission at $2\nu_p$ 10^8 Hz and effective plasma wave temperature of 10^{12} to $10^{13} m_e c^2$. A competition between heating these photons by plasma waves and cooling by electrons (stimulated Comton scattering) produces a steep photon density spectrum $\propto \nu^{-3}$ that is taken over by normal inverse Compton scattering at $\nu_{crit} \simeq 1.5 \times 10^{11}$ Hz. The change of slope produced at this frequency is characteristic of observations and fundamental to the theory. The large infrared emission comes from these photons further heated, times 100, by inverse Compton scattering before escaping from the nebula. Bremsstrahlung and inverse Compton scattering produce an equal X-ray luminoscity. The peculair properties of plasmon-photon scattering and stimulated Compton scattering produce a radio flux $\propto \nu^{-1}$ and a possible explanation of the rapid apparent expansion $v \gg c$ of compact radio sources.

David N. Schramm (ed.), Supernovae. 73-93.

Introduction

There are now three semi-independent reasons to attempt to model the luminosity of quasars using a gravitational **potential well of roughly** 10^8 $M_\odot$. **These** are:

1. The Eddington limit.
2. The total collision rate of stars in a galactic nucleus for a life time $\gtrsim 10^6$ years.
3. The ejected mass flow rate of $\simeq 100$ $M_\odot$/ year necessary to explain line emission behavior for a life time of $\gtrsim 10^6$ years.

We will discuss these at greater length, but _if_ we assume quasars are limited in mass to $\simeq 10^8$ $M_\odot$, then the life time $\simeq 10^6$ years and variability 1 to 10 random events per year (Terrell and Olsen, 1970; Fahlman and Ulryeh, 1975), suggests that individual stars of 10 to 100 $M_\odot$ are involved in the energy release. Of course this need not necessarily be the case, but given the ubiquitous nature of stars, and the evolution rate of galactic nuclei of stars, it would seem that there is some significant motivation to explore possible mechanisms of quasar luminosity based upon energetic bursts evolving 10 to 100 $M_\odot$ of matter. Currently there are several possible mechanisms for producing such outbursts:

1. Stellar collisions in the combined gravitational potential of the stellar system (Spitzer and Saslow, 1966; Gold, Axford, and Ray, 1965).
2. Stellar coalescence and supernovae (Colgate, 1967).
3. Tidal break-up in the gravitational field of a massive black hole (Salpeter, 1964; Lyndon-Bell, 1969; Lyndon-Bell and Rees, 1971; Hills, 1975; and Shields and Wheeler, 1976).
4. Pulsar formation in an assemblage of pulsars (Arons, Kulsrud, and Ostriker, 1975).

Without searching far we will find references to insuperable difficulties with the first three of these mechanisms, and # 4 has not been around long enough, but as the title of this paper would indicate we find a slightly less insuperable difficulty with # 2 than the others. Finally once we have done such a "number" as multiple supernovae in a galactic nucleus of stars, where are we in the still more insuperable problem of producing the bizarre radiation spectrum of a quasar? This last problem we think we can approach with somewhat greater enthusiasm because at least the references proclaiming impossibility are fewer. Arons, Kulsrud, and Ostriker, (1975), have solved this problem provided we form several hundred pulsars per year; we suggest again a different scenario.

Once we have assumed that the energy required for quasar luminosity is released in one to ten outbursts per year in a mass of 10 to 100 $M_\odot$ (a kinetic energy of 1 to 10 Mev/nucleon) and in a gravitational potential well of 10^8 $M_\odot$, then we believe there is a natural progression of physical events that leads to something very close to the observed quasar radio, infrared, optical and X-ray spectrum (Colgate, Colvin, and Petschek, 1975; Petschek, Colgate, and Colvin, 1976).

This radiation mechanism (primarily plasmon-photon emission and stimulated Compton scattering) can probably - but not neccessarily - be applied to the other three energy source mechanisms so that for the purpose of this discussion, some motivation must be given for choosing supernovae. The motivations are:

1. Massive stars are generally accepted progenitors of supernovae and there is reason to suppose that stellar coalescence building massive stars should occur in a 10^8 $M_\odot$ assemblage of stars before evolution proceeds to the sequential stages of disruptive stellar collisions and later a massive black hole.

2. The result of a supernova explosion is generally accepted to be a neutron star. The binding energy of a cold, limiting mass neutron star ($\simeq 2\ M_\odot \times c^2/5$) is significantly greater than that required for the maximum quasar luminosity at 10 per year.

3. Supernovae are popular and have been observed.

There is considerable literature describing the difficulty with all these mechanisms. Sanders (1970) concludes that stellar coalescence leading to supernovae does not occur. The literature attempting to model supernovae is extensive and inconclusive in the sense that theoretically no obvious sequence of events leads unequivocally to a supernovae explosion. Instead we suggest the following scam: collisional coalescence leads to canonical supernovae and low mass garden-variety neutron stars $\simeq 1/2$ to 1 $M_\odot$. The specific binding energy of a neutron star when formed hot is relatively small; this is the basis of the theoretical difficulties in attempting to model explosions. Even when cool, the specific binding energy or 1/2 to 1 $M_\odot$ neutron stars is significantly smaller, namely 0.05 c^2, than the maximal value of 0.1 to 0.2 c^2 at 2 to 2.5 $M_\odot$, assuming a stiff equation of state, obtained by the mean field model (Pandharipande, Pines, and Smith, 1976),, so that if 1 $M_\odot$ could be added to our small neutron star, the available energy is large $\simeq$ 80% of the maximal total binding. We suggest that this happens as the neutron star collides with other stars as well as all stars are doing at this stage of cluster evolution. The mass accretion onto the neutron star would then occur under the ideal circumstances to transfer the incremental binding energy by neutrinos to eject the mantle. This is because the neutron star is initially cold. The other advantage is that the problem is time dependent and three dimensional so that we expect some modest delay before the appearance of new calculations. Also our active neutron star can release its energy in several outbursts before becoming a "modest" black hole. The energy release when the modest black hole collides with further stars leads to even more insuperable difficulties, but it would seem difficult to postulate as large a specific energy release per star swallowed as per star ejected; hence the reason for modeling tidal break-up with a massive black hole, rather than with modest ones. Finally it would appear likely that the explosion would be sufficiently non-spherical that both the mantle and the dense remnant heavier neutron star or modest black hole would be ejected from the galactic nucleus.

The Eddington Limit

We believe the luminosity of the quasar must exceed the Eddington limit so that a steady out-flow of mass occurs as inferred from the interpretation of emission and absorption lines (reviewed in Burbidge and Burbidge, 1967; and in Burbidge and Perry, 1975). This mass flow corresponds to 10 to 100 $M_\odot$ per year during a lifetime of 10^6 to 10^7 years for limiting masses of $5 \times 10^7 \leq M \leq 2 \times 10^9 M_\odot$.

The same emission lines indicate a density $\simeq 10^4$ to 10^8 cm^{-3} and temperature $\simeq 10^4$ degrees such that the opacity will be at least as great as Compton scattering. Hence the Eddington limit becomes:

$$L_{47} = 13 \; M_8 \tag{1}$$

where L_{47} is the luminosity in units of 10^{47} ergs/sec and M_8 is mass in 10^8 $M_\odot$. Contrary to accretion models in black holes, (Salpeter, 1964; Lyndon-Bell, 1969; Lyndon-Bell and Rees, 1971; Hills, 1975; Shields and Wheeler, 1976; Young, Shields, and Wheeler, 1976; Frank and Rees, 1976; and Bahcall and Wolf, 1976) we believe that the luminosity must exceed the Eddington limit. Furthermore it must do so by a considerable factor if the mass out-flow necessary to explain the lines is to be maintained. In several papers Kippenhahn, Mestel, and Perry, (1975); and Mestel, Moore, and Perry, (1976), have developed the theory of radiation driven winds from quasars assuming steady state and consequently ionization equilibria within the non-thermal radiation flux from the central source. These solutions will certainly apply at larger radius than the quasar source surface and especially where the longer time, lower density and radiation flux leads to a steady state. Furthermore at larger radius additional mass in the form of a halo of stars of the central source may increase the gravitational potential from say 10^8 $M_\odot$ to several times 10^9 $M_\odot$ and then the resulting mass flow becomes significantly affected by the gravitational potential. Our problem occurs at the smaller radius inferred from the characteristic fluctuation time of $\simeq$ 1/10 year, giving a light travel time radius of 10^{17} cm. Then in order for this to be defined as a surface, $\int_{R_s}^{\infty} K \rho \, dr \leq 2/3$, and for a near constant velocity v, ρ (at R_s) $\leq 10^{-17}$ g cm^{-3}, and v $\simeq 5 \times 10^9$ cm sec^{-1}. In order to acheive this velocity by the radiation stress of a luminosity of 10^{47} ergs/sec, we assume that at least 1/2 the acceleration must come from the radiation stress or:

$$\frac{L\,K}{4\pi Rc} \gtrsim 1/2\ (v^2/2)$$

or:

$$\frac{L_{47}}{R_{17}} \gtrsim 1 \qquad (2$$

As a consequence, since R_{17} is a near minimum and L_{47} a maximum, any further reduction in the acceleration of matter by gravitational stress will make the inequality more difficult to satisfy and hence the Eddington limit (1) would imply that $M \lesssim 1.3 \times 10^9\ M_\odot$. It is also worth noting that $\dot{M} = 100\ M_\odot$/year is a probable near maximum because the kinetic energy of the mass flow, maintaining the surface condition $\rho KR \lesssim 2/3$, becomes:

$$L_{KE} = \frac{\dot{M}\,v^2}{2} \simeq 10^{46}\ \frac{\dot{M}^3_{100}}{R^2_{17}} \quad \text{ergs sec}^{-1} \qquad (3$$

where $\dot{M}_{100}$ is the mass flux in units of 100 $M_\odot$/year. A slightly smaller value, $\dot{M} < 100\ M_\odot$/year is consistent with the line breadth and emission mechanism of Ptak and Stoner, (1975).

The Evolution Rate of a Galactic Nucleus

Spitzer (1971) reviewed in some detail a possible evolutionary sequence of a galactic nucleus in which he pointed out a combination of two possible routes to the formation of a dense nucleus of stars ($\simeq 10^8\ M_\odot$ in $R \lesssim 1$ pc). The one depends upon stellar evolution and ejection of gas ($\simeq 40\%$ in $\simeq 10^8$ years); the in-falling of the gas; and a new cycle of star formation at smaller ($\simeq 1/10$) radius etc. In addition Spitzer pointed out that stratification of the stellar mass distribution with heavier stars falling towards the center (because of a lower Maxwell equilibrium velocity) becoming isothermal among themselves. These stars lose energy to central crossing low mass stars - thereby evolving to a high density "cusp" of massive stars in a time 1 to 2 relaxation periods (Spitzer and Hart, 1971; Spitzer and Shull, 1975) as compared to a previous canonical evolution time of 88 relaxation periods by "evaporation". The initial concentration from which this process might start can also be formed by cluster capture in the galactic center (Tremaine, Ostriker, and Spitzer, 1975). In Colgate (1967), we pointed out how multiple coalescence among the stars (provided a few were present initially of $M \gtrsim 2\ M_\odot$) could lead to a supernova rate of up to 5 per year. An extensive analysis by Sanders (1970) showed how this estimate was insuperably high and a better maximum rate was 1 per 350 years. This study was designed to show that supernovae were an inadequate energy source for quasars, because the initial stellar mass distribution corresponded to solely 1/2 $M_\odot$ stars and all stars formed from the (large) mass emission rate from non-coalescent collisions were introduced as 1/2 $M_\odot$ stars. It just happens that coalescence occurs roughly a factor of 2 in velocity below that corresponding to $NM \simeq 10^8$ $M_\odot$, and $M \lesssim 2$ $M_\odot$, so that the lack initially of a few larger mass stars $M \gtrsim 2$ $M_\odot$ ensures the result of collisional disruption rather than building to more massive stars. Since the collision time (Spitzer, 1971):

$$\tau_c \propto \frac{M_\odot^{1/2}\, r_\odot^{2}}{M^{1/2}\, r^{2}} \tag{4}$$

and $r \propto M^{3/4}$ one sees that $\tau_c \propto M^{-2}$ and a distribution of more massive stars makes a very large difference in the final question of disruption or coalescence. (A few more massive stars will build by collecting stars to some mass limit $\simeq 60$ $M_\odot$ before a full collision period has occurred.) As Sanders points out the introduction of a distribution of stellar masses "closer to the mass distribution observed in the solar neighborhood", and more especially the central mass stratification expected from dynamical effects both should lead to a significantly different conclusion. The conclusion we still believe is that if the central cusp in stellar density of 10^8 $M_\odot$ consists of 5 to 10% of 2 to 3 $M_\odot$ stars, then the system will undergo coalescence to massive stars 10 to 60 $M_\odot$ and that these will evolve to supernova within the relaxation time of 10^6 to 10^7 years. A calculation similar to Sanders' should be done where the appropriate mass distributions are included. It should also be pointed out that the disruption collisions that follow for $M \simeq 1/2$ $M_\odot$ take place at $\langle v \rangle \simeq 2 \times 10^8$ cm/sec or at 10^{-2} the specific energy density required for producing the quasar luminosity. Hence the stars would have to evolve and accumulate as white dwarfs in order to reach a state where disruptive collisions could occur at great enough energy per event to correspond to the presumed single event fluctuations of quasar luminosity.

The Supernova Scam

Sanders (1970) generously used an integral supernova luminosity of 10^{53} ergs which is x 10 larger than the maximum kinetic energy associated with supernovae and x 10^4 larger than the observed integral luminosity from supernova yet we are presumtive enough to suggest that a supernova of Brand X might emit as kinetic energy one half the maximum star binding energy for a stiff equation of state, i.e., $W_T = M(c^2/5)$ where $M = 2$ $M_\odot$ or where $W_{\text{Brand X}} = 4 \times 10^{53}$ ergs. We suggest that this may happen - not from the first explosion of the 10 to 60 $M_\odot$ stars formed in coalescence, but instead when the neutron star, formed from the near standard supernova, cools - in 10's of seconds - and later, $\simeq 10^6$ years collides with another high mass star.

The vast difficulty in recent supernova calculations (Bruen, 1975; Wilson, 1975;and Wilson et al, 1975) is that the neutrino opacity in the imploding core is really too high to give an explosion. What happens in the calculation is that "neutronization" neutrinos (from protons and electrons being squeezed into neutrons) are actually captured in the implosion to a neutron star and hence neutrino degeneracy maintains, temporarily within the neutron star, a large fraction of electrons and protons, $\simeq 1/3$. The elctrons and neutrino degeneracy pressure remains high and so there is a slow approach to the bound neutron star state as the neutrinos ooze out.

Recently at the Conference on Weak Interactions and Gravitational Collapse, April 1976 at the University of Chicago, a possible resolution of this insuperable dilemma was proposed. Arnett (1976) suggested that the extreme degeneracy of the captured neutronization neutrinos - recognized first by Mazurek (1974), and Sato (1975), but contrary to their cross section treatment- led to a decrease of neutrino effective scattering cross section and hence allowed a very much greater burst of neutrino energy to escape the neutron star at the critical point of maximum collapse - and bounce - and hence led to a large ejection of matter of the envelope. Unfortunately Arnett applied this argument to degenerate neutrinos scattering off of non-degenerate electrons and nucleons. Lamb and Pethick (1976) showed that this did not lead to a change in neutrino scattering cross section and hence no explanation. This was the state of the insuperable difficulties until Baym and Brown pointed out

1. The principal confinement of the neutrinos takes place by scattering on the degenerate nucleons and electrons of the neutron star core. In this circumstance the scattering cross section is reduced by the factor $(kT/E_{Fermi})^2$. This is possibly a large enough factor if the neutron star is cool enough, $kT \leq 5$ Mev compared to the nucleon degeneracy $E_F \simeq 50$ Mev at $\rho \simeq 2 \times 10^{14}$ gm cm^{-3} to allow the sudden escape of the neutron star binding energy at the bounce point.

2. Maxwell, et al. (1975) pointed out that the anti-neutrino generation rate from the pion condensation process should lead to an enhanced cooling of the neutron star matter during collapse because the anti-neutrino cross sections are smaller, $\simeq 1/3$ on electrons and smaller because neutron rich matter has few protons.

Both these effects may lead to a more effective supernova explosion mechanism and possibly we can make a super-supernova from massive stars and partial but not complete mixing by stellar coalescence. It is fun, however, to consider another special supernova - Brand X - that occurs in the peculair circumstances of an evolving galactic nucleus of stars and coalescence.

If we introduce a cold neutron star into a moderately dense ($\rho_{central} \simeq 10^6$ to 10^7 gm cm^{-3}) massive star of 10 to 50 $M_\odot$, then the neutron star is already cold and bound and the often-wished-for standing shock on the neutron core will form (Colgate and White, 1966). The neutrino luminosity of this standing shock can be roughly calculated from the mass accretion rate corresponding to free fall of a solar mass from $\rho = 3 \times 10^6$ gm cm^{-3}, or

$$\tau_{ff} \simeq \left(\frac{1}{3\,G\rho}\right)^{1/2} \simeq 1.4 \text{ seconds, and}$$

$$L_{shock} \simeq \dot{M}\, c^2/5 \simeq \frac{M_\odot}{\tau_{ff}}\, c^2/5 \simeq 2.6 \times 10^{53} \text{ ergs sec}^{-1} \quad (5$$

It has usually been stated (Bruen, 1976; Wilson, 1976) that a neutrino luminosity of 10^{54} ergs sec^{-1} is required for explosion, but this is for the circumstance of the collapse of a dense $\rho_{central} \sim 10^9$ to 10^{10} gm cm^{-3} stellar core. A lower luminosity is required as the surrounding density is decreased. The neutrino flux in an imploding matter field a few mean free paths thick, rapidly distributes the additional binding energy of the core over a region larger than the freely falling matter ($r \simeq 10^7$ cm.) The density of 10^6 to 10^7 gm cm^{-3} is just that expected for a massive star 10 to 50 $M_\odot$ near the end of its stable life during oxygen burning so that there should be a significant fraction of heavy elements up to iron for the enhanced neutral current neutrino scattering to take place.

An example of the conditions become:

1. Neutron star standing shock

$$\dot{M} = \rho_o\, v \times (\text{area})$$

$$\dot{M} = 2 \times 10^{33} \text{ gms in } 1.4 \text{ seconds}$$

$$v \simeq c/3, \text{ area } = 4\pi r^2_{\text{neutron star}} = 1.5 \times 10^{13} \text{ cm}^2$$

$$\rho_o = 9.3 \times 10^9 \text{ gm cm}^{-3}$$

$$\rho_{shock} = 7\rho_o = 6.5 \times 10^{10} \text{ gm cm}^{-3}$$

2. Temperature behind the sh^ck is such that

$\rho_{shock}\, c^2/5 = \frac{29}{8}\, a\, T^4$ including pairs and electron neutrinos, so that $T = 1.2 \times 10^7$ ev.

The electron neutrinos will reach quasi-thermal equilibrium before the muon neutrinos and will be emitted at $\approx 3\, kT_{shock} \simeq 36$ Mev.

$$\sigma_o (E_\nu / m_e c^2)^2 = 9 \times 10^{-41}\ \text{cm}.$$

The thickness of the imploding matter is $\rho_o r_{shock} = 8 \times 10^{15}$ gm cm^{-2} and so, inclusive of partial degeneracy, it will be transparent. The surrounding and imploding stellar core is considerably thicker due to both its mass and the contribution of heavy elements via neutral current cross section. The enhancement of the neutrino scattering cross section is roughly $.45 \frac{\langle A^2 \rangle}{\langle A \rangle} \simeq 10$ and for a $5\, M_\odot$ core at $\langle \rho \rangle = 3 \times 10^6$ gm cm^{-3}, $\rho \sigma_{scat}\, r = 1.7$ which is near optimum for transferring the neutron binding energy to kinetic energy for explosion.

The final insuperable glitch in the picture is how the neutron star is captured and then gets to the center of a massive star. The cross section for capture of the neutron star colliding with a massive one is no different than colliding with a normal star, say a main sequence one, that is considerably denser than the massive one. The criterion for coalescence as discussed by Colgate (1967), and Sanders (1970) is that the mass per unit area be large enough along the collision trajectory to slow down the incident star within one traversal. The collision must take place within a radius such that the gravitational potential of the small star is equal to its kinetic energy. For stars up to $50\, M_\odot$, the density of the envelope is large enough to ensure capture at the average collision velocities of 10^3 km sec^{-1}. Once capture occurs the small star will execute a spiraling orbit rapidly exchanging angular momentum with the larger mass. In the process matter of the small star will be rapidly eroded and mixed with the larger star down to a point of equal densities.

If the erosion and mixing is as efficient as this, then the usual assumption of complete mixing of the larger star is wrong and instead non-evolved stars (hydrogen) will come to rest in non-evolved material and only evolved matter will come to rest - or be mixed with evolved matter of the larger star within approximately the free fall time. The ambient pressure determines the implosion rate onto the neutron star and so the high temperature implosion shock should be turned on only when the neutron star reaches close to the center - otherwise the shock temperature is lower and then the neutrino cross section is too small to capture the emitted neutrino luminosity. The resulting explosion should be asymmetric and a few percent asymmetry should be enough to eject the heavier neutron star or modest black hole out of the galactic nucleus so that further complications are avoided.

The Quasar Luminosity

We have claimed the creation of multiple explosions on a far-out scale, but the form of this explosion energy is a long way from the bizarre radiation spectrum of a quasar. Arons, Kulsrud, and Ostriker (1975) have found one mechanism, non-linear inverse Compton scattering (NIC for those who missed the acronym in their paper), that creates the observed radiation spectrum from the slow wave field of a large number of pulsars and from the relativistic electrons they produce. If one invokes standard pulsars - as they do - then just as in this case if we utilized standard supernovae, one needs 100 events per year to acheive the observed luminosity of a luminous quasar. Instead we have chosen to postulate a Brand X supernova, 20 times larger than the garden-variety, and to reduce the frequency of events to 10 per year. We believe this is closer to both the constraints of galactic nucleus evolution and the observed fluctuations. On the other hand, it is not fundamental to either radiation mechanism and pulsars will certainly add to the commotion. The major and significant advantage of the multiple pulsar model is the direct and qualitatively assured production of the two components that hithertofore have been considered essential for generating quasar radiation, namely magnetic field and highly relativistic electrons for synchrotron emission, and/or NIC radiation. Single massive objects - pulsars, stars, and black holes lack the necessary stochastic variation unless other stars are involved, and then, if other stars are involved, the joint probability of releasing the energy of interaction (star with massive object) highly efficiently in relativistic electrons and in a relatively high magnetic field becomes dubiously small. Hence the multiple pulsar mechanism is a brillant scam. Arons, Kulsrud, and Ostriker do not address the problem of the origin of multiple pulsars, 100/year, other than to admit that the ejected supernova matter must become transparent before the pulsar contributes to the radiation. This time is estimated to be 1 year and is certainly longer than the initial $(\dot{P}/P)^{-1}$. In recognition of this result Bodenheimer and Ostriker (1974) have made a very convincing theory of supernovae based upon the assumption that the primary acceleration of the ejected matter is derived from the pulsar field. Indeed the strong wave acceleration (McMillan, 1950) appears very effective in accelerating matter to cosmic ray energies (Gunn and Ostriker, 1969; Kulsrud, Ostriker, and Gunn, 1972). If charge separation is included (Colgate,1975) the acceleration is still more effective so that the supernova debris may be accelerated for times considerably longer than a year. The ejected matter should gain not just a fraction of the pulsar energy, but at least an order of magnitude more energy than is later emitted as slow waves. This question is crucial not only to the understanding of pulsars and supernovae but also to the choice between the multiple pulsar radiation mechanism and the plasmon-stimulated Compton mechanism for quasar radiation.

The Nebula Plasma

If the stars are evolving in a galactic nucleus, then there will be more collisions that eject some gas from the colliding stars and at least comparable mass will be rubbed off stars as will be involved in coalescing stars. Hence no matter how massive stars collect, resulting in supernovae, there will always be comparable gas emission rubbed off stars that does not initially have escape velocity. Indeed as Spitzer (1971) points out, gas emitted in stellar evolution will fall to the center of the galactic nucleus and initiate a new generation of stars at smaller radius, etc. The question is what happens to this gas when the energy release by supernovae - or what have you - becomes greater than the Eddington limit? As is well known, this leads to accumulating matter bottling up the radiation until extra pressure ejects the additional matter. This condition - as originally discussed in Colgate (1967)- leads to the condition that the optical depth of the nebula has to be such that the characteristic radiation diffusion velocity is greater than the matter ejection velocity.

We have already derived in equation 2 the necessary condition for ejecting 100 $M_\odot$ per year from a radius of 10^{17} cm and yet maintain a surface. This results in a velocity of 5×10^9 cm sec^{-1}, or perhaps half of this for a more reasonable 50 $M_\odot$ per year mass loss. Hence the radiation thickness of the nebula becomes

$$\tau = \int_0^{R_s} \rho K \, dr = \frac{c/3}{V_{ejection}} = 2 \text{ to } 4 \text{ mean free paths}$$

This thickness will later be defined in a slightly more restricted sense as that thickness at which the heating rate of the nebula by the supernova ejecta is no greater than the rate of diffusion of the radiation out. Since ejecta velocity at a kinetic energy $M v^2/2 = 2 M_\odot c^2/10$ where $M = 20 M_\odot$, is $v = 4.5 \times 10^9$ cm sec^{-1}, then $\tau = 2$. Later we will show that the heating rate from the supernova is great enough so that the gas is fully ionized and therefore the opacity is due to the Compton cross section. Hence $n_e \simeq 3 \times 10^7$ cm^{-3} and the mass of gas becomes = 100 $M_\odot$, i.e. of the same order as the mass of Brand X supernova. The radiation spectrum of the quasar is derived from the interaction of the supernova ejected matter with the ionized nebula, and we will next attempt to make plausible how the interaction leads to the peculair non-thermal spectrum of the quasar.

The Non-Thermal Radiation Spectrum

A 50 $M_\odot$ supernova has to expand to a radius of 5×10^{16} cm before it has collided with an equal mass of matter. Ordinarily we would think in terms of shock. Since this thickness is no more than one Compton mean free path, radiation does not have time to reach equilibrium. If this is so, we ask what temperature would occur for a radiation-thin collision at these relative energies and densities.

The dynamic friction between the ions interpenetrating the electron plasma heats the electrons which in turn are cooled by radiation. The most obvious cooling mechanism is bremsstrahlung and if we include approximations for the ion heating and bremsstrahlung cooling rates as discussed in Colgate (1975), Weaver and Chapline (1974), and Weaver (1976) then one finds that the electrons heat to a relatively high temperature:

$$kT_e/m_ec^2 \simeq .635\ E_i^{1/3} \quad (6$$

where E_i is the relative ion energy measured in Mev per nucleon. For our Brand X supernova the ejecta has a kinetic energy of 10 Mev per nucleon, but the relative kinetic energy of the intersecting stream viewed in the center of the mass system at the equal density point is 1/4 of this or 2.5 Mev/nucleon, and $kT_e \simeq m_ec^2$. With this high electron temperature, despite the relative transparency of the nebula, we immediately are concerned with inverse Compton scattering of the bremsstrahlung photons and the resulting enhanced cooling of the plasma. This will ultimately be the source of the X-ray spectra (Lightman, 1974; Katz, 1976), but what concerns us before this , is another source of photons (that will ultimately be Compton scattered) that is far more copious and at lower enrgy. This source is the production and subsequent scattering of transverse photons by longitudinal plasma oscillations. One may ask quite reasonably why seek an explanation in such obscure phenomena when we are so assured of the behavior of incoherent processes as synchrotron emission, bremsstrahlung, and NIC radiation? The motivation arises because of the extraordinary high effective temperature inferred for the radiation processes of a quasar. The fluctuation time defines a size and the red shift and observed flux give a luminosity at a given frequency. If the radiation process is incoherent, this radiation temperature is a lower limit to the energy of the radiating electrons. Since in the radio frequency range - even up to the low infrared, these temperatures are high enough $\gg m_ec^2$ so there is ample motivation to search for coherent mechanisms. The simplest coherent behavior of a plasma is its resonant oscillation at the Langmuir frequency, $\nu_p \simeq 10^4 n_e^{1/2}$. In the laboratory the degree of coherence is not large, but in the peculiar circumstances of the quasar nebula excited by Brand X supernova, an infintesimal excitation of plasma waves represents the most immense departure from thermal equilibrium. In order to recognize how large this coherence or non-thermal behavior can be, imagine thermal electrons in a cubic Debye length. Since only several plasma waves can fit into this box (the three directions) then thermal equilibrium implies an excitation of plasma waves of only several kT. On the other hand the electron thermal energy in the box is $n_e \lambda_D^3 kT_e$ so that <u>if</u> a large amplitude plasma oscillation were excited such that the energy in oscillation were equal to the thermal energy, i.e., $E^2/8\pi \simeq n_e kT_e$ where E is the electric field of the oscillation, then this large amplitude oscillation would be non-thermal or coherent by the factor $n_e \lambda_D^3$. This factor, or plasma parameter, is usually not large but in the high temperature low density conditions of the quasar nebula becomes:

$$n_e \lambda_D^3 \simeq 1.5 \times 10^{17} \left(\frac{kT_e}{m_e c^2} \right)^{3/2} n_e^{-1/2} \quad . \qquad (7$$

This is a large factor indeed; the actual coherence, or effective temperature of oscillation excited by an instability of growth rate Γ/ν_p or equivalently fractional energy content $\epsilon = (E^2/8\pi)/n_e kT_e$ (Drummond, 1962) is

$$kT_{eff} = \pi/2\Lambda(\ln\Lambda)kT_e$$

where $\Lambda \simeq 24\, n_e \lambda_D^3 \epsilon$ so that:

$$\frac{k\,T_{eff}}{m_e c^2} \simeq 1.2 \times 10^{20} \epsilon \left(\frac{k\,T_e}{m_e c^2} \right)^{5/2} n_e^{-1/2} \qquad (8$$

Since the quasar nebula density is $n_e \approx 10^8$, then for modest instability excitation like $\epsilon \simeq 10^{-3}$ and $(kT_e/m_e c^2) \simeq 1/2$ to $1/4$, the effective oscillation temperature becomes $10^{12}\ kT_e$. The counter-streaming ions should maintain at least this order of "weak-turbulence" in plasma oscillations $\epsilon \simeq 10^{-3}$ so that coherent plasma oscillation phenomena should be very prominent.

Plasmon-Photon Emission and Scattering

There are two important coherent processes. These are:

1. Photon emissions at $2\nu_p$ (Aamodt and Drummond, 1964).

2. Absorption and emission of plasmons leading to photon frequency changes $\pm \nu_p$ (Drummond, 1962).

Colgate (1967) attempted to use these processes to explain quasar

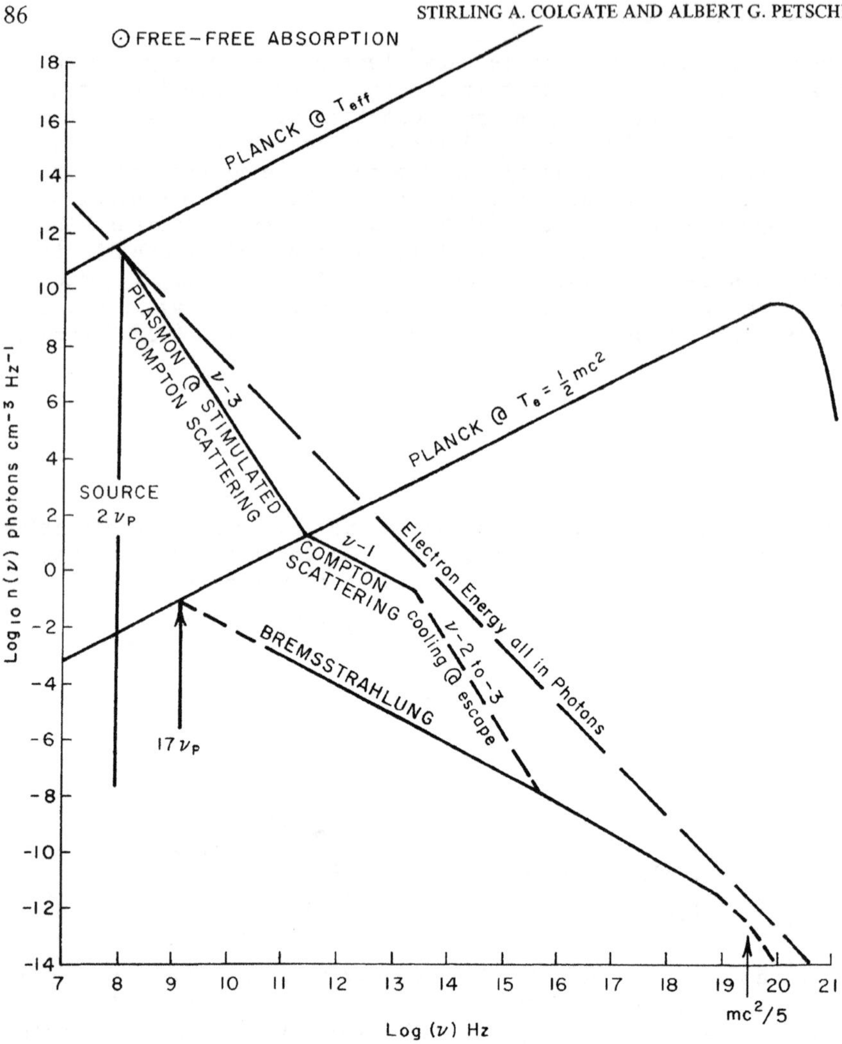

Figure 1. Photon density in the QSO nebula is shown as a function of frequency. Photons are injected by plasma oscillations and diffused by plasmon and stimulated Compton scattering producing the ν^{-3} spectrum until stimulated Compton scattering becomes insignificant near the Planck photon density. Then the spectrum becomes ν^{-1} until escape from the nebula steepens it further. The other main source of photons, bremsstrahlung up-scattered by Compton collisions is also shown, proportional to $\nu^{-1}\ln(kT_e/h\nu)$. This curve is drawn dashed where the spectrum is superluminous. The photon density at $2\nu_p$ is limited by the inverse of the plasmon absorption process at the Planck value of kT_{eff} rather than free-free absorption (⊙). An upper bound on the photon density is given by allocating all the electron thermal energy to photons.

luminosity in the infrared and gummed it up by using a constant cross section for the scattering process 2. Rosenbluth and Lee later did the probelm correctly and, publishing under the pseudonym of Colgate, Lee, and Rosenbluth (1970), demonstrated that the cross section for plasma oscillation photon scattering decreases as $(\nu_p/\nu)^2$ when we consider diffusion in frequency. Since the plasma frequency $\nu_p \simeq 10^8$ Hz, the rapidly falling cross section implied negligible quasar luminosity until Colgate, Colvin, and Petschek (1975) recognized that these up-scattered photons would be a strong source for subsequent inverse Compton scattering. Unfortunately they stupidly forgot to consider stimulated Compton scattering as was pointed out to them by George Chapline of Livermore. Stimulated Compton scattering means that if there are many more photons than in a Planck distribution at the electron temperature, they will heat the electrons rather than vice versa. The stimulated Compton scattering merely means that on the average every collision between an electron and a photon knocks the photon down in energy and the electron up. This disagreeable state of affairs is to be expected because the photon density created by process 1 at $2\nu_p$ frequency is exceedingly large - limited only by the Planck value at kT_{eff}. Therefore rather than obtaining the usual heating effect of inverse Compton scattering, we are faced with a dismal cooling effect. This is all nicely derived in the fundamental paper by Kompaneets (1957), and we have been trying to catch up and understand it ever since. Our latest attempt in this matter (Petschek, Colgate, and Colvin, 1976) actually derives a neat way out of this mess. In figure 1 we show photon number density versus photon frequency. The full Planck spectrum for $kT_{eff} = 1/2\, m_e c^2$ is shown to help with orientation. Photons are injected at $2\nu_p$ by process 1, so copiously that they saturate their Planck limit kT_{eff}. What removes photons from this group - aside from self-absorption - is the stochastic scattering $\pm\nu_p$ or process 2. This is larger than Compton scattering by roughly the coherence factor $kT_{eff}/kT_e \simeq 10^{12}$ so that despite the distasteful cooling by stimulated Compton scattering, the photons diffuse upward in energy against stimulated Compton cooling. What makes for stimulated Compton scattering is the fact that the actual photon number density (and hence effective temperature) is far greater than the Planck photon density associated with the particles, i.e., electrons at kT_e, and hence the photons will undergo a Bose condensation. On the other hand, another population of super particles exists, nonthermal plasma oscillations with a very much higher effective temperature , kT_{eff}, and so the photons are cooled by the one and heated by the other. The proper treatment demands a Boltzman equation diffusion in photon phase space as Kompaneets (1957) has done. However another treatment will illustrate the phenomena. The up-scattering from plasma oscillations is a random walk with a step $\pm\nu_p$ each collision and a cross section

$$\sigma_s \simeq \sigma_{\text{Compton}} \left[\frac{kT_{eff}}{m_e c^2}\right] \left[\frac{\nu_p}{\nu}\right]^2 ,$$

and hence the upward flux in frequency space is:

$$\varphi_{(n)} = - n_e \, c \, \sigma_s \left(\frac{dn_\nu}{d\nu}\right) . \qquad (9$$

The downward flux by stimulated Compton scattering is produced by a downward biased step and because it is stimulated, it is proportional to the photon density squared. Therefore in steady state the fluxes are nearly equal. Hence:

$$-n_e c \, \sigma_s \frac{dn_\nu}{d\nu} \simeq n_e c \, \sigma_c n_\nu^{\,2}$$

and so if we use $\sigma_s = \sigma_c \, (kT_{eff}/ \, m_e c^2) \, (\frac{\nu_p}{\nu})^2$, then $n_\nu \propto \nu^{-3}$. So, in figure 1 we can see how the initial source function at $2\nu_p$ and number density larger than the electron Planck value by the factor kT_{eff}/kT_e decreases as ν^{-3} towards the electron Planck line. Since the Planck distribution increases as ν, the curves intersect at a "critical" frequency

$$\nu_{crit} \simeq (2\nu_p) \times (kT_{eff}/kT_e)^{1/4} \simeq 2 \times 10^3 \, \nu_p$$

An additional factor of 1/4 enters from the more exact treatment giving $\nu_{crit} \simeq 1.5 \times 10^3 \, \nu_p$. We know from the definition of Planck equilibrium that at that photon density the net effect of of the stimulated scattering and the inverse Compton scattering processes balance; in other words at densities below Planck the inverse Compton process dominates; the photons will follow a density behavior determined by inverse Compton scattering. Here the flux from lower frequency is limited and so the upward flux is proportional to both the photon density and to the photon frequency at the Planck line. Inverse Compton scattering multiplies the photon frequency by the factor $(1 + 4kT_e/m_e c^2)$ each average scattering. Hence above the point ν_{crit} where the stimulated curve $(\propto \nu^{-3})$ crosses the Planck value, the photon flux is a constant $n_e \sigma c \nu^2 \frac{dn}{d\nu}$ or $n_\nu \propto \nu^{-1}$. This is similar to the bremsstrahlung spectrum, $n_\nu \propto \nu^{-1} \ln \frac{h\nu}{kT_e}$. After several Compton scattering periods the same spectrum is preserved (Kompaneets, 1957). We would now like to compare the plasmon-photon source to bremsstrahlung. If one equates the bremsstrahlung emission rate to the inverse Compton up-scattering rate neglecting the other photons from plasma oscillations, then the resulting steady state photon number density intersects the corresponding Planck value, i.e., self-absorption would set in, at $\nu_{brem} \simeq 17\,\nu_p$. (In our actual case self-absorption would set in at ν_{crit} because of the high photon number density.) Hence the flux of photons from the plasmon process will be larger than the inverse Compton scattered bremsstrahlung flux at the same frequency ν_{crit} by the ratio:

$$\frac{\phi_{plasma}}{\phi_{brem}} = \frac{(\nu_{crit})^2}{(\nu_{brem})^2 / \ln \dfrac{\nu_{crit}}{\nu_{brem}}} \simeq 7 \times 10^5 \qquad (10$$

It is this very large factor that indicates why plasma processes are important for the generation of low frequency photons.

Inverse Compton Scattered Radiation

The number of collisions required for a photon to escape the nebula, τ mean free paths thick, is on the average $3\tau^2/\pi^2$ for a spherical region and the photon energy increases due to inverse Compton scattering as

$$(1 + 4kT_e/m_ec^2)^{(3\tau^2/\pi^2)} .$$

This is an extremely sensitive function of τ and imposes a stricter physical limit on τ than the previous mass ejection limit. Since the increase in photon energy due to inverse Compton scattering is progressive, if the nebula is too thick, the photons will appear to contain more energy than is available from the supernova ejecta. Hence the heating process must be self-limiting either because it turns off the heating by cooling the electrons or by the photons diffusing out of the nebula. If the electrons are cooled to a significantly lower temperature than 1/4 to 1/2 m_ec^2, then the range of ejecta ions in the high temperature plasma will become significantly smaller than the nebula radius and the momentum of ejecta will efficiently couple to the nebula gas and eject it. Hence both the radiation diffusion time and ejecta energy considerations limit the nebula thickness. We estimate in Petschek, Colgate, and Colvin that the limiting thickness is such that the average photon is multiplied in energy by a factor of x 100 for a Compton thickness of τ = 3 to 5, and kT_e = 1/2 to 1/4 m_ec^2. This means that the peak luminosity in the infrared will extend from 3×10^{11} to 3×10^{13} Hz and above this will fall off in the optical somewhat more steeply than ν^{-1}. Katz (1976) has calculated these spectra in detail and shows how inverse Compton scattering produces the quasar radiation spectrum. He leaves open the question of the reason for the nebula thickness, kT_e, and soft photon source. Similarly the bremsstrahlung spectrum will be inverse Compton heated such that the low energy photons are increased in energy by x 100 but the total X-ray luminosity is not increased by this factor because the Comton cross section falls to less than 1/2 above 100 Kev and a small change in τ allows these photons to escape rapidly. Since the electron temperature is of the order 1/4 to 1/2 $m_e c^2$, premature escape of the Compton heated photons occurs above $\simeq$ 70 Kev. The X-ray luminosity is then reduced below 100 x bremsstrahlung to approximately 6 to 10 times bremsstrahlung (Petschek, 1976). This results in an X-ray luminosity just about the same as the infrared or $L_X \simeq L_{IR} \simeq 5 \times 10^{46}$ ergs sec^{-1} in agreement with the observations (Morgan, et al., 1975; Rieke and Kinman, 1974).

The Emitted Spectrum

The infrared, optical, and X-ray ($\lesssim$ 70 Kev) emitted spectrum should be directly proportional to the photon number density of figure 1, because the Compton cross section is energy independent. At 70 Kev the X-ray spectrum should be somewhat flatter than the normal bremsstrahlung because of the effective escape window at 70 to 100 Kev; above this energy there should be a more rapid fall off than ν^{-1}. The emitted radio spectrum is more uncertain both because of plasmon scattering as well as stimulated scattering. The simplest assumption is that in spherical geometry the perturbation of the escaping flux due to stimulated Compton scattering can be ignored, and then the escaping flux becomes proportional to $(1/\sigma_s)\, n_\nu \propto \nu^{-1}$, with a glitch at ν_{crit} due to the transition from stimulated to direct scattering. This is roughly what is observed for NGC 1068 (Telesco, Harper, and Lowenstein, 1976) although a flatter spectrum is observed for 3 C 273 from 10^8 to 3×10^{10} Hz (Kellerman et al, 1971). However a single isothermal sphere is too simplistic and a multi-layer structure of different temperature and densities is required. There is one factor, however, which is strongly in favor of a plasmon scattering stimulated Compton source, and this is the apparent expanding radio sources.

Expanding Radio Sources

VLB array measurements (Schilizzi, et al, 1975) indicate compact radio sources in quasars expanding at velocities significantly greater than that of light. We know that the effect of scattering is to tend to fill any hole in the distribution function. Therefore if there is a natural hole for photons to escape through, as for instance, a density hole blown by the current supernova or a previous one near the surface, then the photons will tend to stream out the hole preferentially. This is tantamount to forming a poorly defined beam. This beam of radio photons will tend to scatter from plasma oscillations at larger radii and we will see the progressive scattering of the pulse in the beam, say a month in duration, as it progresses outward in the weakly turbulent nebula. If two such pulses are formed, at a reasonable separation of one radian, then the apparent separation rate, as visualized by scattering will be:

$$v_{apparent} = \frac{c \sin \theta/2}{1 - \cos\theta/2} \simeq 8\ c \qquad (11$$

when $\theta = 1$.

Although the details of radio emission are certainly only sketchy at this time, we at least see some peculair advantages to a model based upon stimulated Compton and plasmon scattering.

Summary

We have tried to show how the evolution of a galactic nucleus of stars of the special mass 10^8 $M_\odot$'s leads preferentially to the radiation phenomena of quasars. Radiation flow greater than the Eddington limit leads to the correct mass flow and life time. The evolution of the stellar system leads to collisions and massive stars; rapid supernovae give canonical neutron stars which in turn make much larger supernovae -Brand X supernovae - when a small, cold neutron star coalesces with a massive star. The high energy of ejected matter, 10 Mev per nucleon, and 4×10^{53} ergs, controls the matter density of the nebula of gas to a thickness of $\simeq$ 3 Compton mean free paths and electron temperatures of $\simeq 1/2\ m_ec^2$. These conditions are ideal for plasma wave photon emission at $2\nu_p \simeq 10^8$ Hz and effective plasma wave temperature of 10^{12} to $10^{13}\ m_ec^2$. A competition between heating these photons by plasma waves and coolong by electrons (stimulated Compton scattering) produces a steep photon density spectrum $\propto \nu^{-3}$ that is taken over by normal inverse Compton scattering at $\nu_{crit} \simeq 1.5 \times 10^{11}$ Hz. The change of slope produced at this frequency is characteristic of observations and fundamental to the theory. The infrared emission comes from these photons further heated by x 100 by inverse Compton scattering before escaping from the nebula. Bremsstrahlung and inverse Compton scattering produce equal X-ray luminosity. The peculiar properties of plasmon-photon scattering and stimulated Compton scattering produce a radio flux $\propto \nu^{-1}$ and a possible explanation of the rapid apparent expansion $\nu >> c$ of compact radio sources.

Acknowledgement

So many people have contributed to finding errors in this on-going analysis that it is hard to rame them all, but there have been a great many errors indeed. Foremost among those who have found same are George Chapline, Jeff Colvin, Tom Weaver, Marshall Rosenbluth, and the Ward Volunteer Fire Department. The work has been supported by the Astrophysics Section of the National Science Foundation.

References

Aamodt, R.E., and Drummond, W.E., 1964, Plasma Phys. (J.of Nuc. Energy,Part C), 6,147.

Arnett, W.D.,1976, Neutrino Scattering and Pauli Exclusion in Supernova, pre-print U. of Ill., Urbana, Ill.

Arons,J., Kulsrud, R.K., and Ostriker, J.P., 1975, Ap. J. 198 687.

Bahcall, J.N. and Wolf, R.A., 1976, Ap. J., in press.

Bodenheimer, P., and Ostriker, J.P., 1974, Ap.J. 191 465.

Bruen, S.,1975, Seventh Texas Sym. on Relativistic Astrop., Ann. of the N.Y.Acad. of Science, 262 80.

Bruen, S.,1976, Conf. on Weak Interactions and Stellar Collapse, U. of Chi., Ill., April 6, 1976.

Burbidge, G.R. and Burbidge, E.M., 1967, Quasi-Stellar Objects, W.H.Freeman, San Francisco, 1967.

Burbidge, G., and Perry, J., 1975, On the Masses of the Quasi-Stellar Objects, pre-print MPI-PAE/Astro 74, Dec. 1975, Max Planck Inst. fur Physik und Astrop, 8 Munchen, 40.

Colgate, S.A., 1967, Ap. J. 150 163.

Colgate, S.A., 1975, Ap. J. 195 493.

Colgate, S.A., Colvin, J. and Petschek,A.G., 1975, Ap. J. 197 L 105.

Colgate, S.A., Lee, E.P., and Rosenbluth, M.N., 1970, Ap. J. 162 649.

Colgate, S.A. and White, R.H., 1966, Ap. J. 14 626.

Drummond, W.E.,1962, Phys. Fluids 5 1133.

Fahlman, G.G., and Ulryeh, T.J., 1975, Ap.J. 201 277.

Frank, J., and Rees, M.J., 1976, M.N.R.A.S., in press.

Gold, T., Axford, W.I., and Ray, E.J., 1965, Quasi-Stellar Sources and Gravitational Collapse, ed. J.Robinson, A.I.Sheild, and E.L.Schurhing, U. of Chi., Ill.,p.93.

Gunn, J.E., and Ostriker, J.P., 1969, Phys. Rev. Lett., 22 728

Hills, J.G., 1975, Nature 254 295.

Katz, J.I., 1976, Ap. J. 206 910.

Kellerman, K.I.,Jauncey, D.L.,Cohen,M.H.,Shaffer, B.B.,Clark,B.G., Broderick,J., Ronnang, B.,Rydbeck,O.E.H., Matveyenko,L., Moiseyev, I.,Vitkevitch, V.V.,Cooper, B.F.C.,and Batchelor, R., 1971, Ap. J. 169 1.

Kippenhahn, R., Mestel, L. amd Perry, J.J.,1975, Astron. and Astrop. 44 123.

Kompaneets,A.S.,1957, Soviet Phys. JETP 4 730.

Kulsrud, R.M. Ostriker, J.P., and Gunn, J.E.,1972,Phys.Rev.Lett. 28 636

Lamb, D.Q., Pethick, C.J., 1976, pre-print, Univ. of Chicago, Ill.

Lightman, A.P., 1974, Ap.J. 194 419, 429.

Lyndon-Bell, D.,1969, Nature 223 690.

Lyndon-Bell, D., and Rees, M.J.,1971

Maxwell, O.,Brown,G.E.,Campull,D.K.,Dashen,R.F.,and Manassah,J.T., 1975, pre-print State Univ. of N.Y.,Stoney Brook,N.Y.

Mazurek,T.J.,1974,Nature 252 287.

McMillan, E.M.,1950, Phys. Rev., 79 498.
Mestel, L.,Moore, D.W.,and Perry, J.J.,1976, pre-prints: Expanding Quasar EnvelopesII,Radiation-Driven Winds, Max Planck Inst. fur Physics, 8 Munchen 40.
Morgan,B.,Lampton,M.,Bowyer,S., and Cruddace,R.,1975,Ap.J. 197 25.
Pandharipande,V.R.,Pines,D.,and Smith,R.A.,1976,Ap.J. 208 -
Petschek, A.G.,Colgate,S.A.,and Colvin, J., 1976, Ap.J. in press.
Ptak,R.,and Stoner,R.E.,1975,Ap.J. 185 121.
Rieke, G.H., and Kinman, T.D.,1974, Ap.J. 192 L 115.
Salpeter,E.E.,1964,Ap.J. 140 796.
Sanders,R.H.,1970, Ap.J. 162 791.
Sato,K.,1975,Prog. Theor.Phys., 54 1325.
Schilizzi, R.T,Cohen, M.H.,Romney, J.D.,Shaffer,D.B.,Kellermann, K.I.,Swenson,G.W. Jr., Yen,J.L.,and Rinehart,R.,Ap.J. 201 263.
Shields,G.A.,and Wheeler,J.C.,1976,Astrop. Lett. 17 69.
Spitzer,L.Jr.,1971,Study Week on Nucleii of Galaxies, ed.D.J.K. O'Connell, Pontifica Acad.Scien.,North Holland.Pub.
Spitzer,L.Jr., and Hart, M., 1971, Ap. J. 164 399.
Spitzer,L. and Saslow, W.C., 1966, Ap.J. 143 400.
Spitzer, L.Jr., and Shull, J.M.,1975, Ap.J. 201 773.
Telesco,C.M.,Harper,D.A.,and Loewenstein,R.F.,1976, Ap.J. 203 L53.
Terrell,J.,and Olsen,K.H.,1970, Ap.J. 161 399.
Tremaine, S.D.,Ostriker,J.P.,and Spitzer,L.,1975,Ap.J. 196 407.
Weaver, T., 1976, Ap.J. Supp.Ser., in press.
Weaver, T., and Chapline, G., 1974, Ap.J. 192 L57.
Wilson, J.R.,1975,Neutrino Flow and Gravitational Collapse, pre-print, Varenna Summer School lecture.
Wilson, J.R., 1976, Conf. on Weak Interact. and Stellar Collapse, U. of Chi., Ill., April 6, 1976.
Wilson, J.R.,Couch, R.,Cochran, S.,and Le Blank, J.,1975,Seventh Texas Symp. on Rel.Astrop.,Ann.of the Acad.N.Y, 262 54.
Young,P.J.,Shield,G.A.,and Wheeler,J.C.,1976, pre-print:The Black Tide Model of QSO's,D.of Astr., U. of Texas,Austin.

A PROGRESS REPORT ON SUPERNOVA STATISTICS

G. A. Tammann
Astronomisdes Institut der Universität Basel
European Southern Observatory, Geneva
Hale Observatories, Pasadena

Abstract

The Galactic SN rate is estimated from historical SNe to be $\tau = 15$ (+15, -5) yr. SN frequencies in external galaxies are determined for a distance-limited subsample of Shapley-Ames galaxies and the frequencies are expressed per unit luminosity for different types of galaxies. It is confirmed that the inclination of (Sc) spirals discriminates against the discovery of SNe. The frequency of Type II SNe correlates well with the galaxy color, whereas the frequency of Type I's does not. Absolute magnitudes of sample SNe are determined, and SNe in very late-type galaxies are found to be overluminous. The properties and possible precursors of SNe of Type I and II are briefly discussed. Type II belongs to the very young population and Type I to the interarm population. It is speculated that all SNe may lie in close binaries.

- - -

The success or failure of supernova statistics depends, as with any problem of stellar statistics, on the possibility of defining unbiased samples. In principle, samples are defined in observational astronomy by limits on either distance or apparent magnitude. Of these two alternatives the former kind of sample has in general, statistical properties which are physically more meaningful. However, distance-limited samples are difficult to obtain: they tend either to be (too) small or to be incomplete with respect to the smallest or intrinsically faintest elements.

If one attempts to investigate the statistics of supernovae (SNe) with respect to their parent galaxies an additional difficulty arises: one needs compatible, unbiased samples of supernovae as well as of galaxies.

A third problem arises from the fact that the number of known SNe is still restricted. If in the order of 10^2 SNe can be used for number statistics the accuracy is limited to ± 10 percent. But if the sample is subdivided into different types of SNe and galaxies, one

David N. Schramm (ed.), Supernovae. 95-116.

deals typically with numbers $\lesssim 10$ which implies statistical errors of factors 1.5 to 2.

I. The Frequency of Supernovae in Our Galaxy

Five SNe are known which have occurred in our Galaxy during the last millennium. Although CasA has not been observed visually, these five objects shall be called the "historical SNe" in the following.

Table 1. Historical Supernovae

Supernova	Name	r (kpc)	z (pc)
SN 1006		2.4	600
SN 1054	Crab	2.0	200
SN 1572	Tycho	6.0	150
SN 1604	Kepler	10.0	1200
	Cas A	3.0	110

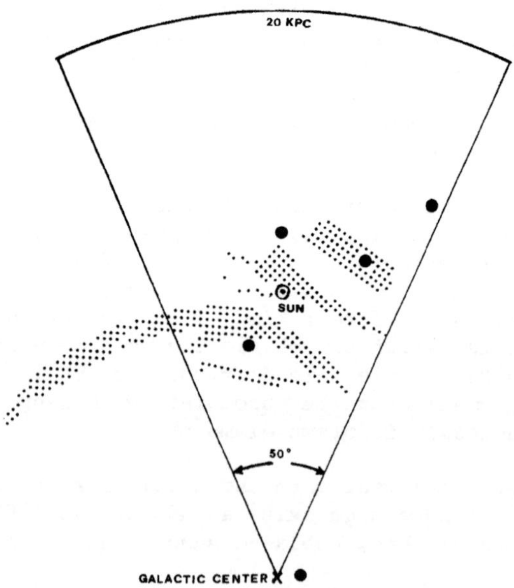

Figure 1. The five historical SNe projected on the Galactic plane. Four SNe lie within a 50° sector; Kepler's SN lies outside.

The historical SNe, listed in Table 1, have known supernova remnants (SNRs) and may be considered as certain cases. Their distances are not accurately known; the values shown should represent the best compromise between different determinations (van den Bergh, 1976). The following conclusions are not sensitive to variations of the adopted distances within factors of ~ 1.5. The distances $|z|$ from the Galactic plane are also shown.

It is imperative to correct the mean apparent time interval between Galactic SN observations of τ = 200 yr for several selection effects to obtain the true Galactic SN frequency:

1. Correction for imcompleteness. The number of n = 5 observed SNe/1000 yr is a lower limit; additional SNe may still be identified. Among other recorded events the one of A.D. 1181 (= 3C58?) may well have been a SN (Clark and Stephenson, 1975; Wilson and Weiler, 1976). Conservatively, a correction factor of k_1 = 1.2 is adopted here, mainly to account for the fact that the historical records fail for $\delta \lesssim -50^{\circ}$ and that some SNe may have been day time events.
2. Correction for surface distribution. Figure 1 shows the historical SNe projected on the Galactic plane. Four objects are confined to a sector with the Sun on its center line and an opening angle of $\alpha = 50^{\circ}$ ($\pm \sim 10^{\circ}$) (Kepler's SN lies near to the Galactic center and outside of α). Hence, the number over the whole plane of the Galaxy is $k_2 n_{50^{\circ}}$ ($n_{50^{\circ}}$ is the number of SNe within 50°), where $k_2 = 360^{\circ}/\alpha = 7.1 \pm 1.0$.
3. Correction for distance. It is not clear whether the 50° sector is actually well covered over its full radial extent. Long visibility is normally a selection criterium for SNe out of the multitude of historically recorded events; in fact, the four visually observed SNe have all been followed for more than one year. During this interval a SN fades on the average by 7^m; this imposes then $m_V(\max) \lesssim -2^m$ and with $\langle M_V \rangle \sim -19^m$ (see below) $(m - M)_{AV} \lesssim 17^m$. Adopting a Galactic absorption of only 1 mag kpc^{-1} yields a distance limit of $\lesssim$ 4 kpc for visually observed SNe of long duration. The distances shown in Table 1 are compatible with this rough estimate (Kepler's SN is clearly an exception). A correction factor $k_3 \geq 1$ should allow for any distance incompleteness within the 50° sector, but lacking quantitative data we assume in the following that k_3 = 1.
4. Correction for z-distribution. In Fig. 2 the z-distribution of Galactic SNRs is shown; the data are taken from Clark and Caswell (1976). The historical SNe, which are not plotted as SNRs, are shown in the upper panel of Fig. 2. The z-distribution of the two samples is drastically different: while the majority of the SNRs lies below $|z|$ = 100 pc, all of the historical SNe lie above this limit. Since there is no reason that their true z-distributions should be different, it must be concluded that either the sample of historical SNe and/or the SNR sample are biased. (The z-distribution changes somewhat with the radial distance R from the Galactic center; the small numbers involved

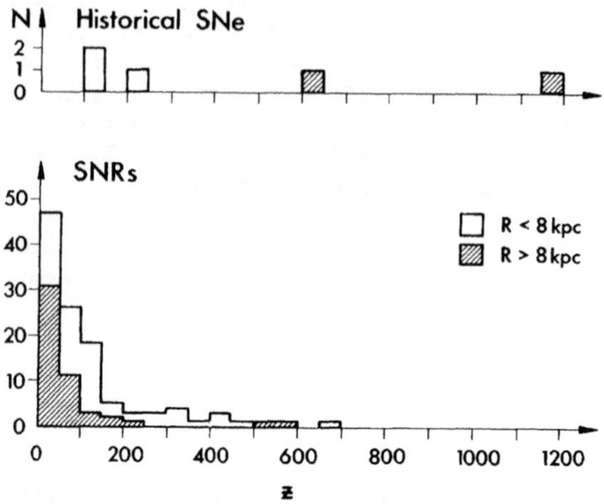

Figure 2. The z-distribution of five historical SNe (upper panel) and of SNRs (lower panel).

do not justify, however, a further subdivision of the sample).

(a) Bias of the historical SNe. It is very possible that SN events at low $|z|$ have been missed due to Galactic absorption. Note that CasA, the historical SN with the lowest $|z|$, has actually not been observed. In that case the true z-distribution of SNe is given by the SNRs. Their distribution can well be represented by

$$N(z) \propto \exp - 0.5\ (z/\sigma_z)^2 \qquad (1)$$

with $\sigma_z \approx 100$ pc. From this follows $N(\text{total})/N(|z| > 100) = 3.2 = k_4$. Thus the $n = 5$ historical SNe with $|z| > 100$ pc should be multiplied with k_4 to obtain the true number of SNe.

(b) Bias of the SNRs. The high - $|z|$ SNRs have shorter life-times than those in the Galactic plane (e.g. Shklovsky, 1974). Hence, the tail of their z-distribution above $|z| > 100$ pc is systematically depleted. In that case, their true σ_z is higher than observed. One knows that SNe are actually in a flattened population (Section Vb) and probably the most extreme, reasonable assumption is that their z-distribution is as broad as that of stars with $\sim 1\mathfrak{M}_\odot$. With $\sigma_z = 300$ pc (Allen, 1973) one then finds from Eq. 1, $N(\text{total})/N(|z| > 100) = 1.3 = k_4$.

It is interesting to look briefly into the statistical evidence for the fast evolution of SNRs at high $|z|$: the last millennium has fed four SNe with $|z| > 100$ pc into the 50° sector. For the same sector Clark and Caswell (1976) list a total of 20 SNRs with $|z| >$ 100 pc. Hence, these SNRs would have a mean age of ~ 2500 yr and account for the last 5000 yr. This time estimate is roughly a factor of three shorter than the new age determinations of typical Galactic SNRs, e.g. the Vela pulsar and the Cygnus Loop (Gorenstein et al., 1974; Kirshner, 1976).

From the discussions under (a) and (b) we shall adopt a mean value of $k_4 = 2.2 \pm 0.7$.

The true number of Galactic SNe during the last millennium becomes then

$$n(1000\ \mathrm{yr}) = (n_{50^o}\, k_1 k_2 + 1) k_4 = 78 \pm 48.$$

(The term 1 in parentheses, for Kepler's SN, should probably be retained here as representative for the innermost SN population.) The error allows for the statistical $\sqrt{n}$ error and for the estimated errors of k_2 and k_4. The mean time interval between Galactic SN events is hence $\tau = 13$ yr with an one-σ range of 8 - 33 yr. This result is in good agreement with an earlier analysis by Katgert and Oort (1967).

The advantage of the present method is that it is independent of any SNR theory, that it is based on a well determined time interval, and that it is fairly insensitive to the adoped distances. The situation is quite opposite if one tries to derive τ from catalogues of SNRs. This is best illustrated by the discrepant results: SNRs give values of τ between 5 yr (Jones, 1975) and 150 yr (Clark and Caswell, 1976).

An independent way to derive the Galactic SN rate is via external galaxies. From the mass of the Galaxy one expects $\tau = 10 - 30$ yr, depending on whether its type is Sc or Sb (Tammann, 1974; hereafter T74). The SN frequencies derived below and a Galactic luminosity of $4 \cdot 10^{10}\ L_\odot$ (Tammann, 1975) give $\tau = 18$ yr if the Galaxy is an Sc spiral, or $\tau \approx 50$ yr if it is of type Sb (as for the relatively large uncertainty for Sb's see below).

The combined evidence from historical SNe and external galaxies confines the probable range of the galactic SN rate to $\tau = 15^{+15}_{-5}$.

Since the historical SNe (except Kepler's) are observed within not too large a distance from the Sun, they should form the best basis to derive the SN density in the solar neighborhood. For the surface density one finds

$$d(\mathrm{pc}^{-2}\ \mathrm{yr}^{-1}) = 0.001\, n_{50^o}\, k_1 k_4 F^{-1},$$

where F is the area within which the SNe have occurred. Two extreme assumptions can be made: (a) the SNe are representative for the whole 50° sector out to R = 15 kpc, then $F = 9.8 \cdot 10^7\ \mathrm{pc}^2$, or (b) the SNe populate a circle with a radius of 6 kpc, then $F = 11.3 \cdot 10^7\ \mathrm{pc}^2$.

Hence, one obtains, always within a statistical uncertainty of a factor of 2,

$$d(pc^{-2}yr^{-1}) = 1.1 \cdot 10^{-10}.$$

This would bring one SN every 200 yr within 4 kpc from the Sun. For the space density of SNe in the solar neighborhood and in the Galactic plane (z = 0) one finds with σ_z = 100 - 300 pc, ρ = (3-9) $\cdot$ 10^{-13} $pc^{-3}yr^{-1}$; the best value is probably $\rho = 5 \cdot 10^{-13}$ $pc^{-3}yr^{-1}$, corresponding - at the present rate of SNe - to about one SN within 3.2 pc from the present location of the Sun during the lifetime of the Galaxy (1.5 $\cdot$ 10^{10} yr).

II. The Frequency of Supernovae in External Galaxies

The following considerations are a continuation of two former papers (Tammann, 1970, hereafter T70; and particularly of T74). A distance-limited sample of galaxies is chosen and the number and type distribution of SNe is studied within this sample. The galaxies and their types are drawn from a revised version of the Shapley-Ames catalogue (Sandage and Tammann, 1976b), which gives redshift information for 97 percent of the galaxies. Only galaxies with $\delta > -36^o$ and $v_o < 1200$ km s^{-1} are considered. The velocity limit corresponds to a distance limit of 22 Mpc (H_o = 55 km s^{-1} Mpc^{-1}). Galaxies with larger velocities ($v_o < 2500$ km s^{-1}), but which are probable members of the Virgo cluster, are also included. Apparent B_T magnitudes of the galaxies are taken from de Vancouleurs and de Vancouleurs (1976, hereafter SCBG); they are corrected for galactic absorption (Sandage, 1973) and the total amount of intrinsic absorption. Since it has been found that the intrinsic absorption corrections by Holmberg (1958) overcorrect the edge-on spirals (Sandage and Tammann, 1976a), a tentative, modified absorption law has been used, limiting the maximum corrections (in the case of edge-on galaxies) to $A_B < 0^m.68$ (for S0/a's), $1^m.33$ (for Sa's to Sb's), $0^m.87$ (for Sbc to Sd's) and $0^m.43$ (for late-type spirals and irregulars). Distances and absolute magnitudes are derived by adopting H_o = 55 throughout. The sample contains 400 galaxies.

The SNe which have occurred in the sample galaxies up to March, 1976, and their types are taken from an updated master list (Kowal and Sargent, 1976). The same modifications have been applied as in T74, giving 76.5 SNe for an adopted search time of 35.3 yr (cf. T74).

The main difference of the present investigation compared to T74 is: (a) a better definition of the sample galaxies due to an increased number of available redshifts, (b) newly assigned galaxies types, (c) the use of B_T instead of m_{pg} magnitudes, and (d) smaller intrinsic absorption corrections (not to face-on orientation, but for the full amount of absorption).

Table 2. Distribution of SNe with Galaxian Type

	E	S0	Sa	Sb	Sc	Ir I	All
n_{Gal}	43	62	37	58	179	18	400
n_{SN}	5	4	-	16	48.5	3	76.5
Type I	5	3	-	6	14	3	31
Type II	-	-	-	6	18	-	24
Others	-	1	-	4	16.5	-	21.5

In T74 several tests were performed for possible biases of the sample. It was found that the section of the sky considered was reasonably uniformly searched, that no distance or SN type discrimination was apparent, and that a finer binning of galaxy types (including the presence or absence of a bar) did not change the results. No evidence could be found that high surface brightness has discriminated against the discovery of SNe close to the galaxy center, but this possibility remains a serious problem. A repetition of these tests does not seem to be warranted here.

It was pointed out in T74 that a distance-limited subsample of Shapley-Ames galaxies is necessarily incomplete as for the more distant dwarf galaxies; this should, however, not affect the present SN statistics.

The linear dependence between the SN rate and the lumonisoty of a galaxy of given type, as derived in T70 and T74, is adopted here as correct.

It should be stressed that the present statistics are based on the observed number of SNe, - the true number is higher. In addition the survey time (35.3 yr) was scaled to the present intensity level of the surveys. Hence, the SN frequencies derived here are to be taken as lower limits.

(a) SNe and galaxy types. The distribution of SNe over the types of the sample galaxies are shown in Table 2. The sample contains only three Ir II (I0) galaxies; they have produced no SN and they are omitted for the rest of this paper. The lack of SNe in Sa galaxies may be a statistical fluctuation; it should be noted, however, that for a much larger sample of SNe only one parent galaxy has been classified as Sa (Maza and van den Bergh, 1976). The occurrence of SNe in Sa galaxies is hence not firmly established yet. As it is well known, Type I SNe are distributed over all types of galaxies, while Type II SNe are confined to Sb and Sc galaxies. In the latter galaxies, the number ratio is Type I: Type II ≈ 1:1 (for face-on Sc's, - probably the least biased sample, - the ratio

Table 3. Dependence of SN Frequencies on Inclination

	n_{Gal}		L (in $10^{10}L_{\odot}$)		n_{SN}		ν_{SN} (in SNu)	
	Sb	Sc	Sb	Sc	Sb	Sc	Sb	Sc
$i \leq 30^{o}$	2	36	5.1	60.4	0	29.5	$\leq$0.6	1.38
$30<i\leq 60^{o}$	32	80	113.9	122.5	8	6	0.20	0.14
$i > 60^{o}$	24	63	103.5	92.2	8	13	0.22	0.40
$i > 30^{o}$	56	143	217.4	214.7	16	19	0.21	0.25

is 1: (2.0 ± 0.8)).

(b) SNe and galaxy inclination. In T74 a very strong effect was found discriminating against the discovery of SNe in inclined spirals. In principle, an observational bias of this kind is expected (because of absorption and high-surface brightness problems), but the size of the effect has been contested by Maza and van den Bergh (1976). In Table 3 the situation is reanalyzed. Sb and Sc galaxies and their SNe from Table 2 are rearranged according to inclination and the SN frequencies, expressed in "SN units" (1 SNu = 1 SN $(10^{10}\ L_{\odot})^{-1}(100\ yr)^{-1}$), are given in the last two columns.

The statistically most reliable result in Table 3 is the frequency in face-on Sc's ($i < 30^{o}$). In inclined Sc galaxies the frequency is lower by a factor of 5.5 ± 1.7, in agreement with T74. It is strange to find here again that the frequency in moderately inclined galaxies ($30 < i < 60^{o}$) is more depressed than the frequency in edge-on galaxies ($i > 60^{o}$). No convincing explanation for this is offered. The data for Sb galaxies do not show the inclination effect, but the insufficient data do not exclude the possibility, that SNe in face-on Sb's are ~ 3 times more frequent than in inclined Sb galaxies. A definite conclusion is hampered by the fact that the distribution of i is not random; there are relatively too many face-on Sc and too few face-on Sb galaxies. It is not clear whether this is a real or an observational effect.

The dependence of SN frequencies on inclination depends, of course, on the adopted magnitude corrections for internal absorption. Since these corrections amount to factors of up to ~ 3 in luminosity, the true luminosity of edge-on galaxies might be over-estimated. However, it is quite well established that spirals do suffer some internal absorption. An indication for the absorption corrections being realistic comes also from the fact that the mean luminosity of a sample Sc galaxy is ~ $1.5 \cdot 10^{10}\ L_{\odot}$ and of an Sb galaxy ~ $4 \cdot 10^{10} L_{\odot}$ independent of the inclination.

(c) SN rates in SNu. SN frequencies (in SNu) are shown in

Table 4. SN Frequencies per Unit Luminosity as a Function of Galaxian Type

	E	S0	Sa	Sb	Sc*	Ir I
L (in $10^{10}L_{\odot}$)	73.0	62.2	90.4	232.4	60.4	6.45
n_{SN} ($n_{Type\ I}$)	5(5)	4(3)	0(0)	16(6)	29.5(6)	3(3)
ν_{SN} (in SNu)	0.19	0.18	($\leqslant$0.03)	(0.19$_5$)	1.38	1.32
$\nu_{Type\ I}$(in SNu)	0.19	0.14	($\leqslant$0.03)	(0.07)	0.28	1.32

* face-on systems only ($i \leqslant 30^{o}$)

Table 4 for different types of galaxies. The lowest frequencies apply to E and S0 galaxies (with the possible exception of Sa's) with a mean frequency of 0.19 SNu and a statistical error of 30 percent. A very bright E/S0 galaxy with $M_B^{o,i} = -22^m$ ($8.8 \cdot 10^{10}L_{\odot}$) is hence expected to have τ = 60 yr. The frequencies in Sc and IrI systems are the same within the errors, i.e. 1.38 ± 0.25 SNu; this is about 7 times higher than in E/S0's. From this follows for a very bright ScI galaxy with $M_B^{o,i} = -21^m.5$ ($5 \cdot 10^{10}L_{\odot}$) τ = 14.5 yr, in very good agreement with the observations (cf. T74, Table IXB).

Kowal (Kowal et al., 1975) has monitored 41 giant Sc galaxies for SNe. The fact that from 1971 to the end of 1974 only one SN (1972q) was found, could be taken as indication that the above frequencies are an over-estimate. However, if one allows for the published search times and for the inclination effect of Sc galaxies, a detailed calculation predicts only 3.2 SNe for the full search interval. Thus the decrepancy is statistically insignificant.

The SN frequency in Sb spirals seems to be as low as in E/S0's; the upper limit for Sa's is even lower. These low values are shown in parentheses in Table 4, because no correction for inclination was applied. The drastic SN frequency difference between Sb's and Sc's seems unlikely. In spite of the somewhat confusing numbers in Table 3, it seems therefore indicated to allow for an inclination effect in Sb's (and Sa's); certainly the evidence from Sc's does favor such a correction. If one allows tentatively for a rather conservative inclination factor of 2.5, the frequencies become 0.5 for Sb's and ~0.1 SNu for Sa's. This has the reasonable consequence that the frequency in Sb's lies between the frequencies of Sc's and E/S0's, and that the frequency of Sa's is not significantly lower than in E/S0's. It should be noted in passing that M31, a very luminous nearby Sb galaxy, has produced one SN (1885a) in 10^2 yr, suggesting a rate of the order of 0.2 SNu. It is to be hoped that the recently discovered SNRs in M31 (Kumar, 1976) shall eventually give a broader basis for the SN statistics in this galaxy.

Table 5. SN Frequencies as a Function of Galaxian Color *

	$(B-V)^{o,i}_T$				
	$\leq 0^m_.40$	$0^m_.40-0^m_.55$	$0^m_.55-0^m_.65$	$0^m_.65-0^m_.80$	$>0^m_.80$
n_{Gal}	48	64	36	46	53
n_{SN}	15	18.5	15	6	9
Type I	6	5	8	3	8
Type II	4	10	5	1	-
Others	5	3.5	2	2	1
L(in $10^{10}L_\odot$)	70.6	169.4	121.0	122.7	105.0
ν_{SN} (in SNu)	0.60	0.31	0.35	0.14	0.24
$\nu_{Type\ I}$	0.24	0.08	0.19	0.07	0.24
$\nu_{Type\ II}$	0.16	0.17	0.12	0.02	-

* I0's (Ir II) are excluded

(d) SNe and galaxy color. Searle et al. (1973) have predicted that if SNe (of Type II) come from massive stars their frequency should strongly correlate with the galaxian color. This prediction was borne out for Type II, but not for Type I SNe (Sargent et al., 1974).

The question is checked here again. For 247 sample galaxies of all types, carrying 86 percent of the total light of the sample, $(B - V)^o_T$ colors are available in the SCBG. They have produced 63.5 SNe, i.e. 83 percent of the same SNe. The galaxy colors were corrected for internal reddening following Holmberg (1958). The galaxies are binned into different color intervals in Table 5. The SN rates for each color bin are also shown; there is some evidence that the SN rate increases with the blueness of the parent galaxy. SNe of Type I, taken separately, do not show the effect. For SNe of Type II the effect is, however, very pronounced. The mean colors of the galaxies within the five columns of Table 5 are $\langle(B-V)^{o,i}_T\rangle$ = 0.32, 0.48, 0.61, 0.73 and $0^m_.87$, respectively. Interpolating and slightly extrapolating the predictions by Searle et al. (1973, Table 2) one expects ratios of the SN rates of 1.00:0.66:0.35:0.26:0.005. The ratios of the Type II rates in Table 5 are: (0.8 ± 0.4): (0.8 ± 0.3): (0.6 ± 0.3): (0.1 ± 0.1): 0. In view of the small-number statistics this does not constitute a sensitive quantitative test of the theory, but the agreement is as good as it can be expected. In any case, a correlation with the expected sign between the rate of Type II SNe and the (B-V) color of the galaxy is established beyond doubt.

III. The Absolute Magnitudes of Supernovae

Many SNe of Type I follow a mean light curve with relatively little scatter (Barbon et al., 1973a). The same seems to hold for Type II SNe, apparently with somewhat larger scatter (Barbon et al., 1974). Mean curves were drawn through the scatter points of the published photographic light curves; they are shown in Fig. 3. The mean pg light curve of Type II was extended mainly by using the SN 1970g in M101 (Barbon et al., 1973b; Bahcall et al., 1976). From published (B-V) color and V magnitude data the corresponding V-light curves were also constructed; they are shown in the lower panel of Fig. 3. Clearly the tails of the light curves depend on only few SNe, and they are less reliable than the earlier parts of the light curves. The light curves of Type I and Type II do not seem fundamentally different, the main difference being the bump of Type II about 50 days after maximum. The size of the bump seems to depend on wavelength. Also the mean decline rates of Type II as well as of Type I are apparently wavelength-dependent. This makes it very improbable that the decline rate should be correlated with the half-life times of any radioactive element. From Fig. 3 it seems quite clear that SNe of Type II also reach an exponential decline (100 - 150 days after maximum) which has been considered by Chiu et al. (1975) as a fundamentally important feature of Type I.

In Fig. 4 the visual light curves of SN 1572 and SN 1604 (Baade, 1945) and of S And (Glyn Jones, 1976) are shown. The mean visual light curves of Fig. 3 are superimposed. For SN 1572 and SN 1604 the mean curve for Type I as well as that for Type II fit the observations equally well until ~ 200 days after maximum. After this interval - taking the tails at face value, - the agreement becomes better for Type I. This could be taken as very weak evidence that SN 1572 and SN 1604 were of Type I. S And was much faster than an average SN of Type I and is in that sense peculiar (Minkowski, 1964).

Apparent maximum magnitudes of SNe, the parent galaxies of which have $v_o < 1200$ kms^{-1}, were compiled from various authors, mainly from Kowal (1976), Barbon et al. (1973a), Sargent et al. (1974) and Rust (1974). The agreement between the various sources is generally quite good. The pg magnitudes (assumed in first approximation to be equal to B magnitudes) were uniformly corrected for galactic absorption, but no correction for absorption within the parent galaxy was applied. Distances of the SNe were calculated from the corrected recession velocity of the parent galaxy and H_o. In view of the quietness of the Hubble flow (Sandage and Tammann, 1975; Tammann, 1976) this should introduce negligible scatter into the adopted distances and into the resulting absolute magnitudes.

The mean absolute magnitudes of Type I and Type II supernovae, grouped according to the type of their parent galaxy, are compiled in Table 6. It should be stressed that the SNe considered here represent a distance - limited sample, whereas most previous authors

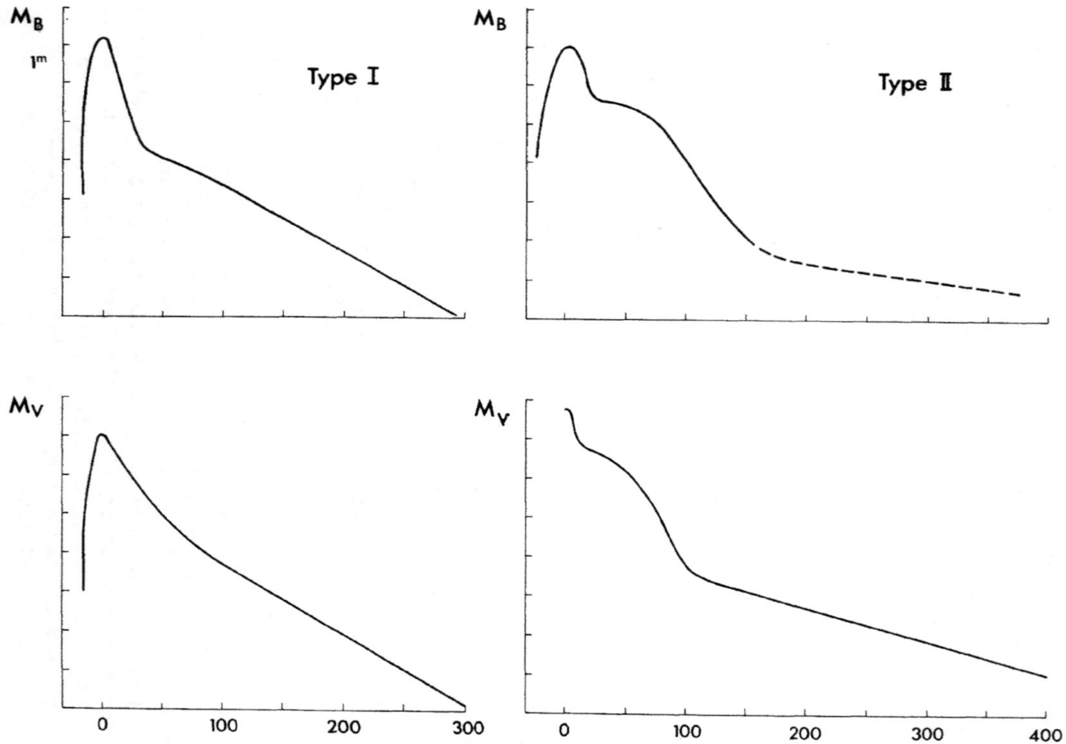

Fig. 3. The average lightcurve in blue and visual light fcr Type I and Type II SNe.

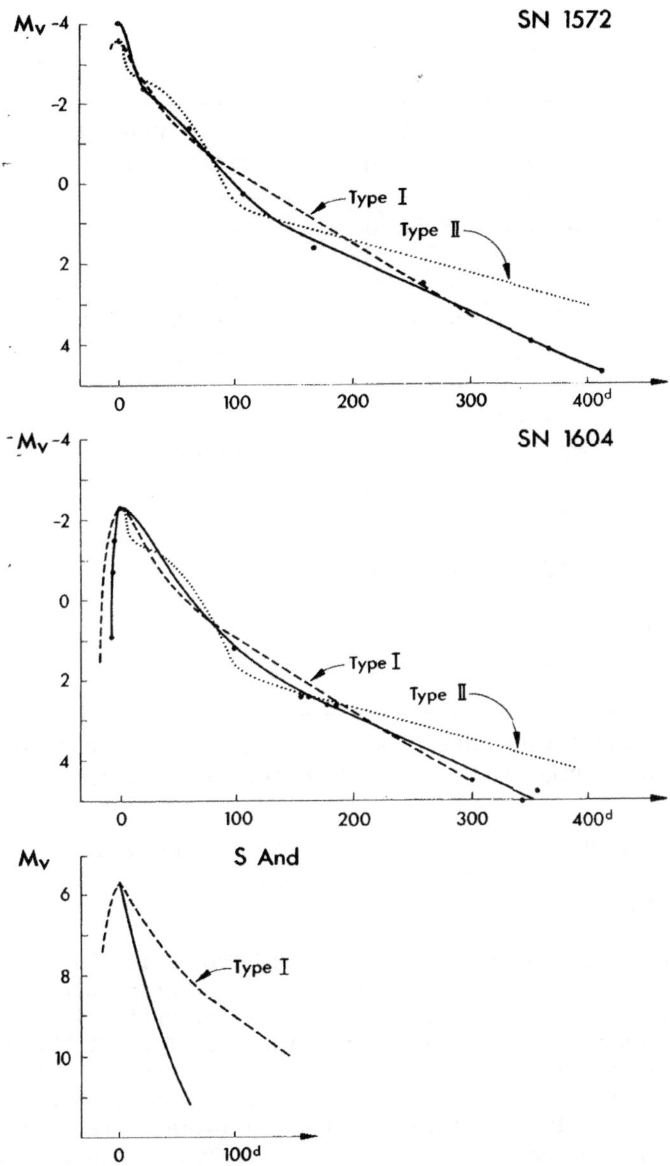

Fig. 4. Visual observations (dots joint by a full line) of SNe 1572, 1604, and SAnd in comparison with the average light curve of Type I (dashed) and Type II (dotted) SNe.

Table 6. Absolute M_{pg} Magnitudes of SNe in a Distance-Limited Sample ($v_o \leq 1200$ km s^{-1}) * (H_o = 55 km $^{-1}$ Mpc^{-1})

	Type I			Type II		
	<M>	σ	n	<M>	σ	n
E, S0, I0	$-19^m.23$ ±.25	$0^m.71$	8	-	-	-
Sa, Sb	-	-	-	$-17^m.73$	-	2
Sc, Sd	-18.31 ±.26	0.70	8	-17.10 ±.22	$0^m.70$	10
Sm, Im	-20.30 ±.22	0.84	4	-	-	-
all	-19.10 ±.25	1.10	20	-17.21 ±.24	0.83	12
all-(Sm, Im)	-18.76 ±.25	0.99	16	-	-	-
Type I slow	-18.83 ±.33	0.80	6			
Type I fast	-19.30 ±.47	1.15	6			

* The absolute magnitudes of a complete, magnitude-limited sample are brighter by the Malmquist correction, $\Delta M = 1.38\sigma^2$.

have studied the absolute magnitudes of SNe in magnitude-limited or unconfined samples. This is the main reason, why some of the present conclusions are new.

The overall mean absolute magnitude of Type I is $1^m.9 \pm 0^m.4$ brighter than for Type II, the scatter about the mean value being somewhat smaller for the latter (!). If the absorption in the parent galaxy is appreciable, it is probably larger for Type II (because of its spiral arm association; Section Vc) than for Type I; this would make the luminosity difference between the two types even smaller. In fact, Minkowski (1964) has proposed quite large corrections for internal absorption: on the average ~ $0^m.6$ for Type I and $1^m.1$ for Type II. But recent spetrophotometric data suggest rather lower absorption values. A detailed study of the absorption in the parent galaxies would be highly desirable.

Barbon et al. (1973a) have divided Type I SNe into "fast" and "slow" subtypes. The data in Table 6 show no significant luminosity difference between these two subtypes.

The relatively large scatter in absolute magnitude of Type I is

Table 7. Absolute Magnitudes of Nearby SNe (Independent of H_o)

	m_V(max)	$(m-M)^o$	A_V	M_V^o (max)	Sources
Crab	-5^m	$11^m.50$	$2^m.14$	$-18^m.64$	1
Tycho	-4	13.89	1.62	-19.51	2, 3, 4
S And	5.7	24.12	0.48	-18.90	5, 6

Sources: (1) Tammann, 1969; (2) van den Bergh, 1975; (3) Goss et al., 1973; (4) Williams, 1973; (5) Glyn Jones, 1976; (6) Sandage and Tammann, 1971.

caused by a pronounced luminosity dependence on the type of the parent galaxy: SNe in very late-type galaxies are apparently brighter by $2^m.0$ than those in Sb/Sc's, SNe in E/S0 being of intermediate luminosity. The formal significance of the luminosity difference is high (6σ). It is unlikely that the full effect could be explained by different amounts of intrinsic absorption in the parent galaxy. For the SN 1972e in NGC 5253, one of the very bright SNe in an Sm galaxy, Chiu et al. (1975) have estimated the total radiative energy output to be $\sim 10^{52}$ ergs; this is probably an upper limit for most other SNe. The variation of the total optical output between the brightest and the faintest SN is about a factor of 50.

For three nearby SNe the absolute visual magnitude can be estimated independent of the value of H_o. The relevant data are compiled in Table 7. Although considerable uncertainties are inherent to these data, the resulting absolute magnitudes scatter by only $\sim 0^m.5$. The mean value of $M_V = -19^m.0$ and $M_B = -19^m.2$ correspond much better to the mean luminosity of Type I than for Type II SNe. This may be taken as additional evidence for SN 1572 being of Type I, and as a suggestion that also SN 1054 and S And belong to this type.

IV. Typology of Supernovae

It is generally agreed that types of SNe can only be determined from spectra. But even with spectra the typing is not a clear-cut procedure. This has led Zwicky to introduce additional types, and quite a number of SNe are typed as peculiar. About 40 days after maximum the spectra of Type I and Type II are so similar, that this has confused at occasions the classification (Rosino, 1976). If it is correct that there are spectra of intermediate type (Fairall, 1972) the distinction of types can hardly be fundamental. In view of the quite similar expansion velocities and black-body temperatures during the early phases, Kirshner et al. (1973) have in fact concluded that "the basic physical situation is rather similar in Type I's and in

Type II's", and they have proposed that a basic difference between the two types lies in the chemical composition of their envelopes.

As discussed in the previous section, there are statistical differences in the light curves and in the absolute magnitude of the two types, but they exhibit considerable overlap: at least one certain Type II SN (SN 1959d in NGC 7331) has a Type I light curve (Arp, 1961), and the brightest Type II's are brighter than the faintest Type I's. It should also be noted that in those galaxies (Sb/Sc's), which have both types of SNe, their frequencies are about equal.

Perhaps more surprising than the statistical correlation of some parameters with SN type is the correlation between properties of SNe of a given type and the type of the parent galaxy. At present there are three such correlations suggested:

(a) Frequency. Type II occurs only in Sb/Sc's as already pointed out. This may just mean that they are very rare in other types of galaxies, and Tinsley (1975) has argued that their rarity is still consistent with the expected death rate of massive stars in these galaxies. For Type I prevails a quite unexpected trend, as already pointed out in T74: their frequency per unit luminosity increases from ellipticals to irregulars by a factor of ~ 7 (± 5). If this effect is real, - and its confirmation would be of great interest, - it would become even more pronounced for the SN rate per unit mass (because the mass-to-light ratio decreases from E's to Ir's) or for the Type I rate per unit luminosity of older population only. Hence, if Type I's come from the older population, the fraction of older stars in Ir's turning into SNe must be higher than in E's.

(b) Light curves. Barbon et al. (1973) have suggested that the "fast" Type I's occur preferentially in later-type galaxies and the "slow" ones tend to concentrate toward the early-type galaxies.

(c) Luminosities. The correlation of the absolute magnitude of Type I SNe with the type of the parent galaxy has been discussed in the previous section (Table 6).

One may conclude from this that Type I SNe in E galaxies tend to be rarer, "faster", and fainter than their counterparts in irregular galaxies, but that those in spirals are the faintest ones.

The main conclusion of this section is that some properties of SNe of a given type vary considerably, and that their variation has the same order of magnitude as the mean difference between Type I and Type II. It is then very difficult to believe that the two types of SNe come from two (or more) kinds of unrelated objects. Instead, it seems tempting to speculate that all SNe are the death mode of one kind of objects, which may have possibly considerable scatter in their initial properties.

V. Supernova Progenitors

So far no pre-SN has been observed. The brightest pre-SN candidate for SN 1970g in M101 has about 21^m; this implies $M \gtrsim -6^m$ and $\mathfrak{M} \lesssim 35\mathfrak{M}_\odot$. It is hoped to derive still a more stringent upper mass limit for this object (Kristian, 1976).

All other information on the SN progenitors has to rest on more indirect evidence. Some information can be gained from the following observations:

(a) z-distribution. The information contained in the z-distribution of Galactic SNRs and historical SNe is fallacious, because it is biased by observational and evolutionary effects, as discussed in Section I, and in addition the objects cannot be separated into Type I and Type II. - The z-distances of SNe can be directly observed in edge-on spirals. Using data on the orientation of individual spirals (with $i > 70^o$) by Nilson (1973) and SN off-set positions from the galaxy center (Kowal and Sargent, 1976) we find for six Type I's $\langle|z|\rangle = 700$ pc and for four Type II's $\langle|z|\rangle = 1000$ pc. These values are clearly upper limits for three reasons: (1) the parent galaxies are not really seen edge-on; (2) the direction of the galaxian major axis, and (3) the off-set positions contain measuring errors which tend to increase $|z|$ sensitively. Better limits on z should become available from a special effort to improve the positional data.

(b) ξ-distribution. If SNe in spirals constituted a spherical population there should be no correlation between their position and the orientation of the major axis of their parent galaxy; on the other hand if SNe belong to the disc population one expects them to concentrate toward the major axis. To check this, those spirals in Nilson's (1973) catalogue, which contain SNe of known type, were selected and the angle ξ determined between the major axis and the radius vector to the SN. The results are plotted in Fig. 5; the data are separated for Type II and Type I and according to the inclination of the parent spiral. It is obvious that the distribution of ξ is not random; particularly in the case of nearly edge-on spirals ($i > 60^o$, hatched histogram) there is a pronounced concentration toward small values of ξ. It is very unlikely that this is a selection effect, because observational problems with the galaxy background would favor SNe near the minor axis of the galaxy and would therefore introduce an opposite trend. Since the effect holds for Type II as well as for Type I there is no question that all SNe belong to the disc population. This is not in contradiction with the occurrence of SNe in elliptical galaxies, because their stellar content resembles the old disc population. The result is in good agreement with the observation that the outer halos of Coma cluster members are SN deficient (per unit luminosity) by at least a factor of 5 (Crane et al., 1976).

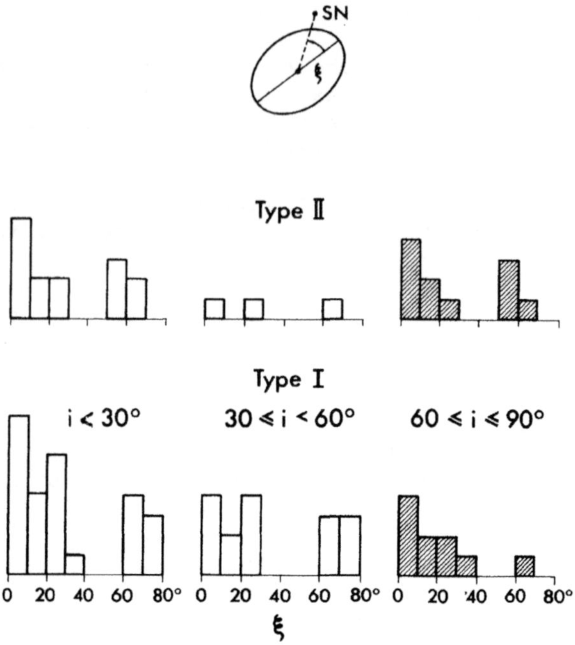

Fig. 5. The angular distance of SNe from the major axis of their parent spiral galaxy. The parent galaxies are divided into face-on, intermediately inclined, and edge-on.

(c) Concentration toward spiral arms. It has been known for a long time that SNe in spiral galaxies are concentrated toward the spiral arms (Reaves, 1953; Johnson and MacLeod, 1963; Moore, 1973). However, it was not clear whether this concerned only Type II's or also Type I's. The recent analysis by Maza and van den Bergh (1976) has revealed quite convincingly that Type I SNe are not associated with the spiral-arm population. Although the interarm regions contain obviously also some massive stars (Schweizer, 1975) it seems from this argument more likely that Type I's do not originate from massive stars, because it is quite doubtful whether there are enough massive interarm stars to roughly balance the frequency of Type I's and Type II's.

(d) Radial distribution. The radial distribution of SNRs (Clark and Caswell, 1976) agrees quite well with the radial distribution of giant HII regions in our Galaxy (Mezger, 1974). The agreement may extend even into the very central regions of the Galaxy where the HII region density is very high. Four recently discovered SNRs within 1° from the Galactic center (Wilson, 1976) suggests that also the SNR density is high in these regions. The good correlation of the radial distributions of giant HII regions and SNRs indicates that at least some SNe belong to the very young population. -The test is not sensitive for all SNe because (1) the sample of SNRs is biased in favor of low $|z|$-values as discussed above, and (2) a contribution of some SNe, -possibly Type I's,- which reflect the radial mass distribution in a spiral galaxy (cf. T74), would still not conceal the particular distribution which is typical for the HII population.

For external spirals Iye and Kodaira (1975) have derived the radial SN distribution separately for Type I and Type II. If the galactic HII distribution is taken as typical for all spirals it can be noted that it agrees quite well with their Type II distribution, and that the agreement is worse for Type I. The absence of the central spike of the Type II distribution might be taken as evidence that some central SNe have not been observed. This, however, would have little influence on the SN frequencies because the areas involved and hence the missing SN numbers are small.

(e) Correlation with galaxy color. The good correlation of the Type II frequency with galaxian color, as discussed in Section IId, is a strong indication for their progenitors being young stars, whereas the lack of such a correlation for Type I's is probably the best evidence that they are not associated with the very young population.

(f) White dwarfs. Weidemann (1968) has estimated the present formation rate of White dwarfs to be (2-5) $\cdot$ 10^{-12} $pc^{-3}yr^{-1}$; this is about four to ten times higher than the SN rate in the solar neighborhood (Section I). If one assumes that about half of the local SNe come from massive stars (Type II) then the most massive $\sim$ 5 to 10 percent of all stars should die as SNe. This is roughly compatible with the observation that stars with $\lesssim$ 4 (or even 6) $\mathfrak{M}_\odot$ can die as WD's (cf. Tinsley, 1975).

Similar mass limits can be determined by comparing the SN rate directly with the death rate of massive stars, derived from some adopted initial mass function. All these methods, however, are only valid <u>if</u> there is a unique mass limit above which all stars turn into SNe (of Type II). The very existence of SNe in the older population (Type I's) questions the correctness of this assumption.

These arguments support the general view that Type II SNe are associated with the very young population (points c, d, and e) and hence that they are the death mode of all or some massive stars. Their occurrence in different types of galaxies is compatible with this assumption (Tinsley, 1975). SNe of Type I belong to the flattened population of spiral galaxies (point b), but to the older, interarm population (points c, e, and d). This makes also their occurrence in E galaxies understandable. The conclusion here concerning the Type I SNe is in variance with T74, where it was concluded that all SNe come from massive stars, but it is in agreement with Tinsley's (1975) results. There is, however, no reason to maintain the old assumption that SNe of Type I were associated with the metal-poor, spherical Population II.

VI. Conclusions

The main conclusion is a negative one: we are still lacking a basic piece of understanding about the SN precursors. Why are there two types of SNe, which obviously belong to different populations, and which in spite of that show amazing overlap of their observed properties? Why, if Type I SNe are associated with the old or older population, do their properties apparently vary with the galaxy type?

Either SNe come from two entirely different kinds of progenitors, or there is only one type of SN precursors, which then must have a considerable spread in mass and age. In the first case, one has to assume the existence of a funnel mechanism, which forces the initially different conditions into a quite narrowly defined state just prior to the explosion. In the second case, one could think of objects which are not confined to a special population, for instance of close binaries. Several authors have proposed close binaries as the precursors of Type I SNe (e.g. Hartwick, 1972; Whelan and Iben, 1973) and it is well known that for stars with masses $\gtrsim 5\,\mathfrak{M}_\odot$, - the probable pregenitors of Type II, - "the occurrence of a single star is a rather rare phenomenon" (Blaauw, 1961). The common property of all SNe would then be that they are in binaries, in which probably mass exchange would set the proper stage for the SN explosion. In that case, the variation of the properties of Type I's with the galaxy type would not be too surprising, because it is not implausible that the frequency and average properties of binaries change with the type of the galaxy.

At this point this is, of course, a mere speculation, and a result on the actual physics of the SN phenomenon cannot be expected from a statistical analysis.

The author would like to thank Drs. S. van den Bergh, C.T. Kowal, J. Kristian, L. Rosino, A. Sandage, L. Searle, V. Weidemann, T.L. Wilson, and L. Woltjer for very stimulating and helpful discussions. Miss B. Flach and Mrs. R. Kraan have given very valuable help in the data processing. Without the initiative of Dr. D.N. Schramm and the support of the Swiss National Science Foundation this paper would not have been written.

References

Allen, C.W. 1973, Astrophysical Quantities, Third Edition, London: Athlone, p. 251.
Arp, H. 1961, Ap. J. 133, 883.

Baade, W. 1945, Ap. J. 102, 309.
Bahcall, J.N., Kowal, C.T., and Tammann, G.A. 1976, to be published.
Barbon, R., Ciatti, F., and Rosino, L. 1973a, Astron. Astrophys. 25, 241.
Barbon, R., Ciatti, F., and Rosino, L. 1973b, Astron. Astrophys. 29, 57.
Barbon, R., Ciatti, F., and Rosino, L. 1974, Supernovae and Supernova Remnants, ed. C.B. Cosmovici, Dordrecht: Riedel, p. 115.
Bergh, S. van den 1975, Galaxies and the Universe, ed. A. and M. Sandage and J. Kristian (= Stars and Stellar Systems, vol. IX), p. 509.
Bergh, S. van den 1976, private communication.
Blaauw, A. 1961, B.A.N. 15, 265.
Chiu, B.C., Morrison, P., and Sartori, L. 1975, Ap. J. 198, 617.
Clark, D.H., and Caswell, J.L. 1976, M.N. 174, 267.
Clark, D.H., and Stephenson, F.R. 1975, Observatory 95, 190.
Crane, P., Tammann, G.A., and Woltjer, L. 1976, in press.
Fairall, A.P. 1972, M.N.A.S.S.A. 31, 23.
Glyn Jones, K. 1976, J. Hist. Astron. 7, 27.
Gorenstein, P., Harnden, F., and Tucker, W. 1974, Ap. J. 192, 661.
Goss, W.M., Schwarz, U.J., and Wesselius, P.R. 1973, Astron. Astrophys. 28, 305.
Hartwick, F.D.A. 1972, Nature Phys. Sci. 237, 137.
Holmberg, E. 1958, Medd. Lund Obs., Ser. II, No. 136.
Iye, M., and Kodaira, K. 1975, Publ. Astron. Soc. Japan 27.
Johnson, H.M., and MacLeod, J.M. 1963, P.A.S.P. 75, 123.
Jones, E.M. 1975, Ap. J. 201, 377.
Katgert, P., and Oort, J.H. 1967, B.A.N. 19, 239.
Kirshner, R.P. 1976, Ann. Rep. Director Hale Obs. 1975-76, in press.
Kowal, C.T. 1976, private communication.
Kowal, C.T., and Sargent, W.L.W. 1976, private communication.
Kowal, C.T., Sargent, W.L.W., and Huchra, 1975, P.A.S.P. 87, 401.
Kristian, J. 1976, private communication.
Kumar, C.K. 1976, P.A.S.P. 88, 323.
Maza, J., and Bergh, S. van den 1976, Ap. J. 204, 519.
Mezger, P.G. 1974, The Interstellar Medium, ed. K. Pinkau, Dordrecht: Reidel, p. 9.
Minkowski, R. 1964, Ann. Rev. Astron. Astrophys. 2, 247.
Moore, E. 1973, P.A.S.P. 85, 564.
Nilson, P. 1973, Uppsala General Catalogue of Galaxies.
Reaves, G. 1953, P.A.S.P. 65, 242.
Rosino, L. 1976, private communication.
Rust, B.W. 1974, Thesis, University of Illinois.
Sandage, A. 1973, Ap. J. 183, 711.
Sandage, A., and Tammann, G.A. 1971, Ap. J. 167, 293.
Sandage, A., and Tammann, G.A. 1975, Ap. J. 196, 313.
Sandage, A., and Tammann, G.A. 1976a, Ap. J., in press.
Sandage, A., and Tammann, G.A. 1976b, in preparation.
Sargent, W.L.W., Searle, L., and Kowal, C.T. 1974, Supernovae and Supernova Remnants, ed. C.B. Cosmovici, Dordrecht: Reidel, p. 33.
Schweizer, F. 1975, La dynamique des galaxies spirales, ed. L.

Weliachew, Paris: CNRS, p. 337.
Searle, L., Sargent, W.L.W., and Bagnuolo, W.G. 1973, Ap. J. 179, 427.
Shklovsky, I.S. 1974, Sov. Astr. 18, 1.
Tammann, G.A. 1969, Mitt. Astron. Ges. 27, 55.
Tammann, G.A. 1970, Astron. Astrophys. 8, 458 (T70).
Tammann, G.A., 1974, Supernovae and Supernova Remnants, ed. C.B. Cosmovici, Dordrecht: Reidel, p. 155 (T74).
Tammann, G.A. 1975, Optische Beobachtungsprogramme zur galaktschen Struktur und Dynamik, ed. Th. Schmidt-Kaler, Bochum, p. 1.
Tammann, G.A. 1976, Redshifts and the Expansion of the Universe, ed. C. Balkowski and B.E. Westerlund (= I.A.U. Coll. No. 37), in press.
Tinsley, B.M. 1975, P.A.S.P. 87, 837.
Vancouleurs, G. de, and Vancouleurs, A. de 1976, Second Reference Catalogue of Bright Galaxies, Austin: University of Texas Press (SCBG).
Weidemann, V. 1968, Ann. Rev. Astron. Astrophys. 6, 351.
Whelan, J., and Iben, I. 1973, Ap. J. 186, 1007.
Williams, D.R.W. 1973, Astron. Astrophys. 28, 309.
Wilson, A.S., and Weiler, K.W. 1976, Astron. Astrophys. 49, 357.
Wilson, T.L. 1976, private communication.

MASSES OF SUPERNOVA PROGENITORS

Beatrice M. Tinsley
Yale University Observatory

1. INTRODUCTION

What stars become supernovae? To the observer, this question asks for the prior identities of stars that initiate recorded supernova (SN) outbursts. To the theoretician, this question asks for the nature of stellar models that explode with ejection of matter and a suitable flash of light. Aiming to strengthen the rather tenuous connections between these views, I shall discuss here how a variety of empirical clues bear upon some unsolved theoretical problems concerning the origins of SN. Resolution of these problems is of interest, because SN have been implicated in a wide range of astrophysical contexts, including nucleosynthesis, the origin of cosmic rays, dynamics of the interstellar medium, and the origin of pulsars and black holes.

Theories and observations are discussed extensively by other authors in these proceedings, and in many earlier reviews (e.g., Minkowski, 1964; Woltjer, 1972; Oke and Searle, 1974; and articles in the books edited by Schramm and Arnett, 1973; Tayler and Hesser, 1973; and Cosmovici, 1974). The issues may be defined by the following brief outline. Most well-observed SN fall into two classes, which are defined on the basis of light curves and spectra, although standard theories of their origins depend primarily on their sites within galaxies. Type II supernovae (SN II) are exclusively associated with young stellar populations in spiral galaxies, so they are thought to be explosive deaths of massive stars. Type I supernovae (SN I) occur in all types of galaxies, including ellipticals, so they are assumed to arise from very old stars, such as exploding white dwarfs. However, as discussed below, the sites of SN I do not unambiguously prove that all (or any) have extremely old precursors, even though it is accepted that the dominant stellar population of elliptical galaxies is old. Alternative scenarios postulate that types I and II differ chiefly in the absence or presence, respectively, of a hydrogen-rich stellar envelope, as indicated by their spectra.

The problems to be addressed here may be introduced in the context

David N. Schramm (ed.), Supernovae. 117-129. *All Rights Reserved.*

of theoretical ideas on how stars die - that is, how they behave after their normal nuclear burning stages as giants and supergiants.

A single star less massive than $1.4M_{\odot}$ cannot burn carbon, so it dies as a white dwarf. This remnant could later explode if it gains too much mass from a companion or if long-term instabilities operate; these processes have been invoked as the origin of SN I because they can occur billions of years after the initiating star (evolving companion or white dwarf progenitor) formed, and thus are possible in galaxies where star formation ceased long ago. A white dwarf origin for SN I involves some theoretical problems, such as the lack of a model for the explosion itself, and difficulties in accounting for type I light curves if the envelope mass is much less than $1M_{\odot}$ (G. Lasher, these proceedings).

Stars initially more massive than $1.4M_{\odot}$, up to a limit to be denoted m_w, also die as white dwarfs because mass ejection via a red giant wind and planetary nebula reduces them below $1.4M_{\odot}$ before the stage of carbon ignition.

If mass loss is efficient enough for m_w to exceed the limit (denoted m_c and theoretically $\sim 8M_{\odot}$) above which the core is non-degenerate on carbon ignition, more massive stars undergo successive stages of nuclear burning until the core is exhausted, collapses, and (by an unknown mechanism) causes explosive ejection of the overlying matter. This process is the standard scenario for SN II. It predicts the formation of a neutron star or black hole condensed remnant, and the release of stellar nucleosynthesis products in proportions that bear a promising resemblance to the solar system abundance distribution. It has recently been suggested (Renzini, 1976) that SN I could have similar origins, arising from slightly less massive precursors than SN II so that the explosion occurs after the hydrogen-rich envelope has been lost in a stellar wind.

If some stars initially less massive than m_c do not shed enough mass to become white dwarfs (i.e., if $m_c > m_w$), carbon ignition is thought to lead to detonation, which may disrupt the entire star and eject disproportionate amounts of iron-peak elements. This "carbon detonation SN" is not identified with any empirical SN type, and the hope is that the process is prevented by mass loss to $1.4M_{\odot}$ before the carbon ignition stage in all stars with degenerate cores on the giant branch; that is, carbon detonation will not occur if $m_w \geq m_c$.

Stars in interacting binary systems could die differently. On one hand, rapid mass loss could lead to premature death; the possible significance of this process for SN rates, nucleosynthesis yields, etc., deserves investigation. On the other hand, excessive mass gain could lead to instability of, for example, a white dwarf; this process is invoked in standard theories of the origin of novae and of SN I.

Two outstanding questions will be emphasized in this paper.

I. <u>What is the lower mass limit for stars to die explosively?</u> (Delayed explosion of a white dwarf and unusual effects in close binaries are not included here.) In particular, we wish to know whether the upper mass* limit m_w for death as a white dwarf exceeds the upper mass limit m_c for potential carbon detonation, so that in fact detonation does not occur.

II. <u>What stars initiate type I supernovae?</u> It is especially interesting to ask whether SN I have progenitors with masses less than m_w (as in the exploding white dwarf scenarios, single or binary), or greater than m_w (as in the scenarios where they differ from SN II mainly in envelope properties alone). These alternatives will be referred to as "low-mass" and "massive" progenitors, respectively. This question is of theoretical interest, for understanding the fate of white dwarfs and alternative SN I progenitors, and, as will be discussed, it carries important implications for the history of star formation in elliptical galaxies.

The types of evidence to be considered here are local SN rates, empirical estimates of m_w, the distributions of SN among stellar populations in galaxies, and colors of SN-producing galaxies. The following discussion is based on an earlier review (Tinsley, 1975), which contains further details and references. Other sources include the reviews by Tammann (1974) on SN statistics, Woolf (1973) on non-explosive stellar mass loss, and Trimble (1975) and Audouze and Tinsley (1976) on various aspects of stellar and galactic evolution.

2. LOCAL SUPERNOVA RATES AND DISTRIBUTIONS

The SN rate at the sun's distance from the Galactic center can be inferred from the numbers and ages of pulsars - presumably the stellar remnants of at least some SN - and of expanding interstellar supernova remnants (SNR). It is convenient to begin by supposing that the SN responsible for these remnants arise from all stars with masses greater than some value m. Thus we define a function d(m), which is the deathrate (expressed in units of no./pc^2/y) of stars more massive than m; for short-lived stars ($m \gtrsim 2M_\odot$), d(m) is identical to the corresponding birthrate function, which is inferred from counts and theoretical lifetimes of O and B stars on the main sequence. Recent estimates, for the relevant mass range $3 < m/M_\odot < 15$, can be expressed as the value of m corresponding to a given deathrate, in the form:

$$\log m \simeq -0.47 \log d - 4.3 \ (\pm 0.1), \qquad (1)$$

where m is in $M_\odot$ and d is in no./pc^2/y (cf. Figure 1 of Tinsley [1975], which is based on Schmidt [1963] and Ostriker <u>et al</u>. [1974]). In

---

* Stellar masses refer to their initial main sequence values, unless otherwise stated.

spite of the uncertainties, the slope in this mass range appears to be shallower (i.e., the birthrate as a function of mass is steeper) than Salpeter's (1955) average slope; the difference is significant in the context of supernova progenitors.

We can now use estimates of the local SN rate, equated to d(m), to derive a lower mass limit m for stars to die as SN. It is useful to note that main-sequence O stars are thought to have (initial) masses of about $10M_{\odot}$ and more, early B stars about $5 - 10M_{\odot}$, and late B stars about $2 - 5M_{\odot}$.

2.1 Supernova remnants

Various estimates, based on radio and historical data, have in the past led to time intervals between SN explosions in the Galaxy of $\tau_{SNR} \simeq 50 \pm 25$ y (Tammann, 1974). A significant recent development is the use of X-ray data to evaluate shock temperatures in SNR, and hence to determine their ratios of explosion energy to ambient gas density, which in turn calibrates the radio SNR count-diameter relations in terms of SNR formation rates. By this method, Clark and Culhane (1976) have estimated $\tau_{SNR} \simeq 110 - 190$ y. Now the Galactic distribution of SNR derived by Ilovaisky and Lequeux (1972) relates τ_{SNR} to a local rate in the units used for d(m), which can be inserted in equation (1) for a mass denoted m_{SNR}. This mass is interpreted as the lower limit for stars whose SN lead to SNR, on the assumption that SNR do indeed arise from stars of a given mass and greater. Corresponding rates and mass values are given by the following table:

τ_{SNR} (y)	50	100	150	200
Rate ($10^{-11} pc^{-2} y^{-1}$)	4	2	1.3	1.0
m_{SNR} ($M_{\odot}$)	3.9	5.4	6.3	7.4

These numbers are consistent with the view that SNR arise from stars that were OB types on the main sequence, as is also suggested by their flat Galactic distribution. There are too many uncertainties to draw detailed conclusions from the estimates of m_{SNR}, or to discuss the possibility of an upper mass limit for SNR formation.

Some important problems of principle set limits on the usefulness of extended remnants as probes of SN progenitors:

i) Presumably, SN of all types give rise to SNR (Kepler's supernova providing direct evidence that SN I can do so), so if only SN II arise from massive stars, their precursor mass range must be less extensive than indicated by the total SN frequency.

ii) The lifetimes of SNR scale approximately as $n^{-1/2}$, where n is the ambient interstellar gas density, so there is a strong bias against finding remnants of SN that exploded far from the Galactic plane. It may even be that the agreement of m_{SNR} with the masses of OB stars merely reflects the fact that OB stars form and die within the gas layer, so there is no information on the possibility of some less massive precursors. We might therefore expect the total Galactic SN rate to exceed the rate of formation of observed SNR.

2.2 Pulsars

Pulsar precursors were first discussed on the basis of their distribution and lifetimes by Gunn and Ostriker (1970). Most recently, the sensitive survey by Hulse and Taylor (1975) has been analysed in these terms by J. H. Taylor and R. N. Manchester (in preparation). The latter authors derive a formation rate in the solar neighborhood of $r_p \simeq 10^{-11}\ fT_7^{-1}/pc^2/y$, and a mean Galactic interval between births of $\tau_p \simeq 60f^{-1}T_7$ y, where f is the incompleteness factor due to the beam size of pulsar radiation and T_7 is the mean pulsar lifetime in 10^7 y. (These estimates include several selection and incompleteness effects not considered by Roberts [1976], who derived a much smaller rate from the same data.)

The values of f and T_7 are clearly crucial to the interpretation of these results. Taylor and Manchester prefer f = 5 and $T_7 = 0.2$, a remarkably short lifetime suggested by proper motion data. These values lead to $r_p \simeq 2.5\times10^{-10}$ and $\tau_p \simeq 2$y. The implication is that only a small fraction of pulsars are associated with events that lead to an extended SNR, although the total SN rate in the Galaxy could be as high as one in 2 y, according to the statistics in § 2.3. This high pulsar rate, inserted in equation (1), implies a lower mass limit $\sim 2M_\odot$, which is implausible and suggests that many pulsars may have low-mass precursors (e.g. possibly being associated with SN I, if these originate from white dwarfs).

Problems with the high rate can be avoided if one takes f = 1 and a lifetime $T_7 \simeq 1$, as indicated by spin-down timescales. Then $r_p \simeq 10^{-11}/pc^2/y$ and $\tau_p \simeq 60$y, in line with the values for SNR.

Further details of pulsar beam sizes and lifetimes are needed. As discussed by Gunn and Ostriker (1970), the mass range for pulsar precursors cannot be inferred directly from their distribution perpendicular to the Galactic plane, because they evidently have acquired anomalous velocities, perhaps in a kick at birth.

2.3 Analogy with external galaxies

The Galactic SN rate can be estimated from SN statistics for similar external galaxies. Tammann (1974) has derived in this way a mean time interval between SN in the Galaxy of $\tau_{SN} = 10\ (+4.2,-2.3)$y if the Galaxy is an Sc, or $\tau_{SN} = 31(+40,-11)$y if it is an Sb. These intervals are directly proportional to the masses adopted for external galaxies, so they scale inversely with Hubble's constant, for which Tammann used the value H_o = 55 km/s/Mpc. Estimates of τ_{SN} based on galactic luminosities scale as the inverse square of H_o. In view of the evidence that H_o is no smaller than 55 (e.g., de Vaucouleurs, 1976; Peebles, 1976; Bottinelli and Gouguenheim, 1976), the above values are if anything overestimates of τ_{SN} from external galaxies. It thus appears that the total Galactic SN rate is several times the rate of formation of SNR. A possible explanation, already mentioned above, is that a large fraction of Galactic SN explode too far from the gas layer to leave long-lived SNR. If so, many SN may not have massive precursors, coming instead from lower-mass stars with scale heights greater than those of OB stars and gas; such stars do not have identical birth- and deathrates, so an equation like (1) above cannot readily be used to estimate their mass range.

As yet, the uncertainties in statistics of external SN and in selection effects possibly reducing the SNR rate below its true value are too great to allow one to conclude whether or not all Galactic SN arise from OB stars. The current evidence suggests that many SN do indeed arise from stars whose explosions do not lead to extended SNR, so perhaps their precursors are of low mass.

3. THE PROGENITORS OF WHITE DWARFS

Another approach to finding the minimum mass for stars that die explosively is to estimate the maximum mass, m_w, for death as a white dwarf. (Later explosion of the white dwarf is a separate consideration - here we are addressing question I of § 1.)

3.1 White dwarfs with companions

White dwarfs associated with main-sequence stars in clusters or in binary systems presumably arose from stars more massive than the living companions. The star Sirius B and the numerous white dwarfs in the Hyades and Praesepe clusters immediately give the limit $m_w > 2M_\odot$. Since planetary nebulae are presumably the immediate precursors of white dwarfs, the existence of A and late B type companions in a few cases (Woolf, 1973) points to $m_w > 2.5M_\odot$. These data already indicate a significant amount of mass loss, because the white dwarf progenitors must have reduced to $1.4M_\odot$ (in fact, significantly less, according to white dwarf mass estimates).

More information can be obtained from clusters by comparing the numbers of white dwarfs with an extrapolated main-sequence mass function. In this way, van den Heuvel (1975) estimated that the white dwarfs in the Hyades arose from stars with masses up to 3 - $6M_\odot$, with a best estimate $m_w = 4M_\odot$. A similar range of progenitor masses was derived by Sweeney (1976) from theoretical white dwarf cooling times.

The probable membership of a normal white dwarf in the Pleiades cluster (Jones, 1973; Greenstein, 1974; and references therein) points to an interestingly large value, $m_w \gtrsim 6M_\odot$, which lies within the theoretical uncertainty limits for m_c. It would be unwise to base strong conclusions on just one star; one worry is that the cluster could perhaps contain some stars much older than those at the early B star turnoff. Searches for white dwarfs in other clusters with B and early A star turnoffs are underway (J. R. Angel, private communication), and the results of these will provide much firmer limits to m_w.

3.2 Stellar mass loss rates

Stars with initial masses up to about $4 - 6M_\odot$ appear to have strong enough winds as red giants to reduce below $1.4M_\odot$ before reaching the stage of carbon ignition (Fusi-Pecci and Renzini, 1976; Mengel, 1976; Scalo, 1976). The implied value of m_w is thus in agreement with the above estimates from white dwarfs. An accurate value of m_w cannot yet be derived by this method because of important uncertainties, especially in the stellar mass loss rates.

On the question of whether m_w exceeds m_c, Mengel (1976) points out that core degeneracy itself leads to large red giant luminosities, and so drives rapid mass loss, with the result that stars below m_c may directly protect themselves from carbon ignition. In other words, the values of m_w and m_c may be constrained to be equal.

In summary, white dwarfs and stellar mass loss rates provide a consistent case for stars dying quietly if their initial masses are below m_c, and becoming SN at a later stage of nuclear evolution if their initial masses exceed $m_w \geq m_c \sim 6M_\odot$. However, the conclusion against carbon detonation is far from firm, for m_w could be as small as $4M_\odot$ and m_c could be as great as $8M_\odot$.

4. POSITIONS OF SUPERNOVAE AMONG STELLAR POPULATIONS IN GALAXIES

The distributions of SN in galaxies of different types provide very important clues as to their precursor masses. The following discussion is based largely on the statistical studies by Tammann (1974) and Maza and van den Bergh (1976); further references are given by those authors and by Tinsley (1975), and Tammann's work will be updated in his paper in these proceedings.

4.1 Supernovae in spiral galaxies

SN II are clearly associated with extreme population I stars. They have been identified only in galaxies with strong spiral structure, and they occur mainly within the spiral arms themselves. The precursors of SN II are thus almost certainly the O and early B stars that define spiral arms by their associated HII regions. A lower mass limit $\gtrsim 6M_\odot$ is indicated; if the limit were smaller, most SN II would occur outside spiral arms, because the slope of the mass function (eq. [1]) means that the majority of stellar deaths come from stars near the lower limit. (The occasional SN II outside a spiral arm could arise from a "runaway" star, as pointed out by W. D. Arnett [private communication]; or, if only the existence of a hydrogen-rich envelope distinguishes SN II from SN I, unusually slow mass loss could be the cause.)

SN I cannot be so easily associated with a stellar population type in spiral galaxies. They are not predominantly among true (halo) population II stars, as Tammann has shown from their bias against discovery in inclined spirals. But they are not members of the very young population I, because they are not associated with spiral arms (Maza and van den Bergh, 1976); therefore, at least most SN I have precursor masses below $\sim 6 - 8M_\odot$. This information is consistent with either alternative mass range for SN I: they could have precursors below m_w, or in some mass range say $\sim 4 - 8M_\odot$ above m_w.

4.2 Supernovae in elliptical galaxies

The most telling clues on the origin of SN I may lie in the elliptical galaxies that have produced SN. If white dwarfs are the origin of SN I, they could occur in regions of galaxies with no evolving stars above about $1M_\odot$, i.e. with the very old stellar population that clearly dominates the light of elliptical galaxies. But if SN I have massive precursors ($> m_w$), all SN-producing elliptical galaxies should be currently making stars. This seems unlikely at first sight, for at least some of the relevant galaxies (e.g., M87, the site of SN 1919a) show no tell-tale signs such as nebulosity at the position of the recorded SN outburst.

The case is not clear-cut, however, because we know from spiral galaxies that SN I are not actually associated with HII regions. If star formation were restricted to late B types and later, it would be possible to have SN I from "massive" precursors, between say m_w and $8M_\odot$, yet without HII regions to indicate that stars are forming. Two very recent pieces of evidence give tentative support to this hypothesis: (1) A. D. Code reported at I.A.U. Symposium No. 74 (August 1976, Cambridge) that M87 itself has unusually strong ultraviolet radiation for an elliptical galaxy; the upturn at short wavelengths is consistent with radiation from early-type stars. Unfortunately, there is not enough spatial resolution to say whether such stars could be present at the position of the SN 1919a. (2) S. van den Bergh has presented evidence at the current I.A.U. General Assembly that the "Sombrero" galaxy, NGC 4594, has blue knots with no Hα emission, most readily interpreted as sites of formation of later B stars, with no accompanying O or early B stars. These pieces of evidence obviously still leave open the possibility that SN I have low-mass precursors, but they suggest that alternative scenarios deserve careful consideration.

It is intriguing to note that some elliptical galaxies do have HII regions and OB stars; notable examples are NGC 5128 (Blanco et al., 1975) and NGC 5253, which has produced two SN in spite of its small size. In addition, large amounts of neutral hydrogen have recently been discovered in E and SO galaxies (Krumm and Salpeter, 1976; Gallagher et al., 1976). The presence of HII regions leads to the prediction that elliptical galaxies should produce an occasional SN II. Frustratingly, only 20% of the SN in E/SO galaxies on Maza and van den Bergh's (1976) list are classified: there are 8 SN I, 32 unclassified. The plea for observational details, for all possible future SN, bears repeating.

5. COLORS OF GALAXIES WITH SUPERNOVAE

It is well known that the colors of galaxies become bluer along the Hubble sequence from ellipticals to late spirals and irregulars (Ir I), and that the trend is due mainly to the increasing importance in the light of massive blue stars. The SN rate (per unit galaxy luminosity) should therefore be greater in blue galaxies than in red ones, although this prediction must be qualified because the integrated colors of a galaxy depend on other parameters (e.g., metallicity, stellar birthrate function, etc.) in addition to its relative content of massive stars.

We can compare the SN rates in galaxies with different colors with relative stellar birthrates in model galaxies with those colors; to the extent that SN come from massive stars, whose lives are short enough for their birth- and deathrates to be essentially equal, the model birthrates should be proportional to the SN rates - subject to the qualifications mentioned above. I shall summarize the results of such studies by Searle et al. (1973), Sargent et al. (1974), Larson and Tinsley (1974), and Tinsley (1975, with results from additional unpublished models).

The relative deathrates of massive stars, predicted by models, are usefully expressed in terms of a parameter R_B, the ratio of total present stellar birthrate to integrated blue luminosity. The ratio between R_B and the model SN rate per unit galaxy luminosity depends on the mass range of SN precursors and on the initial mass function for massive stars. We consider series of models in which the initial mass function and galaxy age are held fixed, but the time dependence of the stellar birthrate varies. Along series with decreasing ratios of present to total past stellar birthrate, R_B of course declines and the color becomes redder; in models redder than B-V $\sim$ 0.9, even the blue light is dominated by very old stars, so R_B is extremely small.

This behavior is reflected qualitatively in the rather meager statistics for colors of galaxies with SN II (Table VII of Sargent et al. [1974]), and in the distribution of SN II with galaxy type (Table 4 of Larson and Tinsley [1974]). The agreement provides further support for massive precursors for SN II, and some support (which is weak because of the paucity of data) for assumptions such as "universality" of the initial mass function that were used in the models.

When all types of SN are considered, the rate does not drop sharply for redder galaxies, again suggesting that at least some SN I arise in older stellar populations whose deathrate does not scale with R_B. If galaxies are subdivided by Hubble type rather than by color, there is indeed a relatively low total SN rate in E/SO's, which would be consistent with their SN having massive precursors due to a small but finite residual star formation rate; whether the rate is small or zero cannot be decided from integrated colors available to date (Larson and Tinsley, 1974) - detailed studies of individual SN sites, as discussed in § 4 above, or more powerful colors, as discussed below, are needed.

The foregoing statistics are based on B-V colors because of their availability for many galaxies. But the interpretation of R_B-(B-V) relations is confused by the problem that the V, B, and even U light of most galaxies has only a minor contribution from the OB stars whose deathrate is of interest. The value of R_B predicted for given colors depends sensitively on the initial mass function adopted for stars not sampled in the colors, and on the functional form adopted for the past stellar birthrate. The considerable scatter of galaxies in the U-B, B-V plane shows that B-V alone is not an adequate index of the stellar birthrate. This point is illustrated by the spiral galaxy M83 (NGC 5236), which has produced four SN and whose U-B color is $\sim$ 0.2 mag bluer than average for its B-V (cf. Figure 2 of Tinsley [1975], based on data from de Vaucouleurs and de Vaucouleurs [1964, 1972]). Such colors are most readily explained as due to an intense burst of star formation, at several times the average current rate for that galaxy or for others with its B-V. Evidently, we should not expect a very tight correlation between B-V and SN rate. We need more direct criteria than UBV colors for estimating the birthrates of massive stars in galaxies.

Potentially useful indices include the strength of Hα emission (Cohen, 1976) and the ratio of radio flux to blue luminosity (Biermann, 1976). The Hα index and the thermal part of the radio flux (or any other index of the number of ionizing photons) reveal only the formation of stars hot enough to form HII regions; this restriction may cause problems in the interpretation of SN statistics, since later B stars could be in the range of "massive" SN precursors and there is suggested evidence (van den Bergh's interpretation of the blue knots in NGC 4594, discussed above) that such stars can form without O and early B stars. Another problem is that non-thermal radiation from SNR themselves could contribute much of the 5 GHz radiation (Biermann, 1976), so the radio-to-blue luminosity ratio does not clearly test the relation between star formation and SN rates.

Colors at satellite ultraviolet wavelengths are very promising.

For example, they could distinguish between elliptical galaxies having no current star formation and those forming stars fast enough to account for the average SN I rate, even though the UBV colors differ very little between the two cases (Larson and Tinsley, 1974).

To summarize, the presently available evidence from galaxy colors does little more than confirm the conclusions drawn from the distributions of SN among stellar populations: SN II rates are (within the considerable uncertainties) proportional to stellar birthrates inferred from galaxy colors, as expected of SN with massive precursors. But if SN I also have massive precursors, the accompanying star formation activity has yet to be detected in many of their parent galaxies.

6. CONCLUSIONS

The evidence discussed above supports conventional answers to the two questions posed in § I, but it has not provided unambiguous answers to the important theoretical problems.

I. The dividing line between death as a white dwarf and death by SN (apart from possible delayed explosion of the white dwarf) appears to lie between 4 and $8M_{\odot}$, right in the theoretical uncertainty range. If carbon detonation remains a problem for nucleosynthesis, we can take comfort in the fact that there is no compelling statistical evidence <u>for</u> its occurrence: stars with degenerate helium cores could lose enough mass to become white dwarfs without carbon ignition. Further research on the occurrence of white dwarfs in open clusters and on mass loss by red giants could be very informative. Estimates of rates suggest that not all SN leave extended remnants, so SNR statistics do not provide values for the precursor masses of most SN. Inferences from pulsars depend critically on uncertain lifetimes and beam sizes; it is possible that not all pulsars correspond to SN with massive precursors.

II. Type I SN sometimes occur in regions with no trace of a young stellar population, so low-mass precursors are indicated. However, known properties of E/SO galaxies by no means exclude a rate of star formation that would account for their SN by massive precursors, especially if the stellar mass spectrum sometimes excludes O and early B types. Type II SN are expected on rare occasions in E/SO galaxies, because some contain HII regions; therefore the question of SN I precursors should be approached by detailed studies of the sites of supernovae unambiguously classified as Type I.

ACKNOWLEDGMENTS

I wish to thank J. H. Taylor for unpublished results on pulsars, numerous astronomers for discussions, and the Aspen Center for Physics and the Institute of Astronomy, University of Cambridge, for hospitality. This research was supported in part by the Alfred P. Sloan Foundation and the National Science Foundation, Grant AST75-16329.

REFERENCES

Audouze, J. and Tinsley, B.M.: 1976, Ann. Rev. Astron. Astrophys. (in press).
Biermann, P.: 1976, Astron. Astrophys. (in press).
Blanco, V.M., Graham, J.A., Lasker, B.M., and Osmer, P.S.: 1975, Astrophys. J. (Letters) 198, L63.
Bottinelli, L. and Gouguenheim, L.: 1976 (preprint).
Clark, D.H. and Culhane, J.L.: 1976, Monthly Notices Roy. Astron. Soc. 175, 573.
Cohen, J.G.: 1976, Astrophys. J. 203, 587.
Cosmovici, C.B.: 1974 (ed.), 'Supernovae and Supernova Remnants', D. Reidel Publishing Company, Dordrecht, Holland.
Fusi-Pecci, F. and Renzini, A.: 1976, Astron. Astrophys. 46, 447.
Gallagher, J.G., Knapp, G.R., Balick, B., and Faber, S.M.: 1976 (in preparation).
Greenstein, J.L.: 1974, Astron. J. 79, 964.
Gunn, J.E. and Ostriker, J.P.: 1970, Astrophys. J. 160, 979.
Huevel, E.P.J. van den: 1975, Astrophys. J. (Letters) 196, L121.
Hulse, R.A. and Taylor, J.H.: 1975, Astrophys. J. (Letters) 201, L55.
Ilovaisky, S.A. and Lequeux, J.: 1972, Astron. Astrophys. 18, 169.
Jones, B.F.: 1973, Astron. Astrophys. Suppl. 9, 313.
Krumm, N. and Salpeter, E.E.: 1976, Astrophys. J. (Letters) 208, L7.
Larson, R.B. and Tinsley, B.M.: 1974, Astrophys. J. 192, 293.
Maza, J. and van den Bergh, S.: 1976, Astrophys. J. 204, 519.
Mengel, J.G.: 1976, Astron. Astrophys. 48, 83.
Minkowski, R.: 1964, Ann. Rev. Astron. Astrophys. 2, 247.
Oke, J.B. and Searle, L.: 1974, Ann. Rev. Astron. Astrophys. 12, 315.
Ostriker, J.P., Richstone, D.O., and Thuan, T.X.: 1974, Astrophys. J. (Letters) 188, L87.
Peebles, P.J.E.: 1976, Astrophys. J. 205, 318.
Renzini, A.: 1976, in R.J. Dickens and J.E. Perry (eds.), 'The Galaxy and the Local Group', R.G.O. Bull. No. 182 (in press).
Roberts, D.H.: 1976, Astrophys. J. (Letters) 205, L29.
Salpeter, E.E.: 1955, Astrophys. J. 121, 161.
Sargent, W.L.W., Searle, L., and Kowal, C.: 1974, in C.B. Cosmovici (ed.), 'Supernovae and Supernova Remnants', D. Reidel Publishing Company, Dordrecht, Holland, p. 33.
Scalo, J.M.: 1976, Astrophys. J. 206, 215.
Schmidt, M.: 1963, Astrophys. J. 137, 758.
Schramm, D.N. and Arnett, W.D.: 1973 (eds.), 'Explosive Nucleosynthesis', The University of Texas Press, Austin.
Searle, L., Sargent, W.L.W., and Bagnuolo, W.G.: 1973, Astrophys. J. 179, 427.
Sweeney, M.A.: 1976, Astron. Astrophys. 49, 375.

Tammann, G.A.: 1974, in C.B. Cosmovici (ed.), 'Supernovae and Supernova Remnants', D. Reidel Publishing Company, Dordrecht, Holland, p. 155.
Tayler, R.J. and Hesser, J.E.: 1973 (eds.), 'Late Stages of Stellar Evolution', IAU Symposium No. 66, D. Reidel Publishing Company, Dordrecht, Holland.
Tinsley, B.M.: 1975, Publ. Astron. Soc. Pacific **87**, 837.
Trimble, V.: 1975, Revs. Mod. Phys. **47**, 877.
Vaucouleurs, G. de: 1976, Astrophys. J. **205**, 13.
Vaucouleurs, G. de and Vaucouleurs, A. de: 1964, 'Reference Catalogue of Bright Galaxies', The University of Texas Press, Austin.
Vaucouleurs, G. de and Vaucouleurs, A. de: 1972, Mem. Roy. Astron. Soc. **77**, 1.
Woltjer, L.: 1972, Ann. Rev. Astron. Astrophys. **10**, 129.
Woolf, N.J.: 1973, in R.J. Tayler and J.E. Hesser (eds.), 'Late Stages of Stellar Evolution', IAU Symposium No. 66, D. Reidel Publishing Company, Dordrecht, Holland, p. 43.

EVOLUTION OF SUPERNOVA PROGENITORS (SNP)

Z. Barkat
Racah Institute of Physics, The Hebrew University of Jerusalem, Jerusalem, Israel

§1. INTRODUCTION

Our aim in this paper is to briefly review the current state of research on the present subject and to try to identify some of the weaker points.

For the purpose of completeness some if not most of our talk will be a repetition of earlier reports on a similar subject (i.e. Imshennik and Nadyozhin 1973, Massevitch and Tutokov 1973, Arnett 1975, Barkat 1975). We shall not deal particularly with the very important subject of SNP binaries. This subject will certainly become a central one in the near future. It is, however, quite clear that the basis for this subject is the work on SNP in the case of single stars to which we address ourselves here. Also, not within our talk range will be effects of rotation and magnetic fields -- this does not mean that they are necessarily negligible. Already at this point we note that we shall not compare evolution "with" and "without" neutrino losses. We rather concentrate on the "with" case only. Some comparisons can be found in the works mentioned before.

Let us start by reminding the reader that most of the work on final stages of stellar evolution has been a study of evolution of stellar cores whose mass stays fixed. Hydrogen envelopes and hydrogen burning shells are in such work either assumed to be absent or at least to have negligible effect on the structure and evolution of the core. In some cases even the Helium envelope was stripped off (carbon-oxygen, C-O cores). Besides the problems which this research itself has been facing (and we shall touch upon these shortly) one of course must check whether and when such a core (bare, and fixed mass), can indeed be regarded as representing the true evolution of stellar cores. It is this second part of the problem that has attracted more attention recently.

David N. Schramm (ed.), Supernovae. 131-141.

§2. FIXED MASS CORES

Figure 1 gives a representation of well-known results of study of the evolution of fixed mass C-O cores. Here the agreement between various authors (Vila 1966, 1967; Rakavy et al 1967; Savedoff et al 1969; Ikeuchi et al 1971, 1972; Boozer et al 1973; Paczynski 1971; Arnett 1972a, b, 1974a, b, 1975a; Kovetz et al 1975; Reiss et Barkat, 1976) is generally quite good even though there are differences in detail (see later). The essential results common to all authors are:

Subdivision into 4 mass regions for which the final stage is qualitatively different. These are:

Region I -- very massive cores ($M_c \geq 30M_\odot$) which encounter a dynamical instability due to pair formations just before central oxygen ignition.

Region II -- massive cores ($2<M_c/M_\odot \leq 30$) which burn successive fuels and then develop iron (nickel) cores and encounter a dynamical instability due to the photodisintegration of nuclei at high temperature ($T \geq 5x10^{9°}k$).

Region III -- intermediate cores ($1.44 \leq M_c/M_\odot \leq 2$). In this relatively narrow region cores evolution is finally dominated by electron capture on suitable nuclei and becomes dynamic. Some of these may even burn silicon while others hardly burn oxygen.

Region IV -- low mass cores ($M_c/M_\odot \leq 1.44$). These cores die peacefully as degenerate white dwarfs. The more massive of these cores will burn their carbon. In figure 1 we give, for example, the ignition line of carbon and it is obvious that there is a minimal core mass (M^C_{min}) for carbon ignition (for which corresponds also maximal ignition density $\simeq 10^7 g/cm^3$). Minor differences between authors exist regarding the exact value of M^C_{min}, (which depends on the imput physics) but it is close to $1M_\odot$. It is also well-known that for the less massive cores just above M^C_{min}, ignition occurs first off center. Central ignition occurs for $M \simeq 1.15M_\odot$ (Boozer et al 1973). Similarly, only cores more massive than $M \simeq 1.39M_\odot$ will ignite oxygen (again off center, for the less massive cores).

Some of the problems which research has been facing here are on the computational side but most are more on the physical side. Improvement in recent years has mainly been an inclusion of much better nuclear reaction networks (Arnett 1974a, b, 1975). The major problems are still associated with the treatment of convection: what is the correct criterion for convective instability (Ledoux or Schwartzschild), can semiconvection be important during these later stages, how far can overshooting be important (Shaviv and Salpeter 1973, Maeder 1975, Castelani et al 1971). Not less important is the treatment of burning shells: structure, stability (Ikeuchi et al 1971, 1972; Kovetz et al 1975), formation and evolution of associated convective zones. We shall come back to some of these questions later.

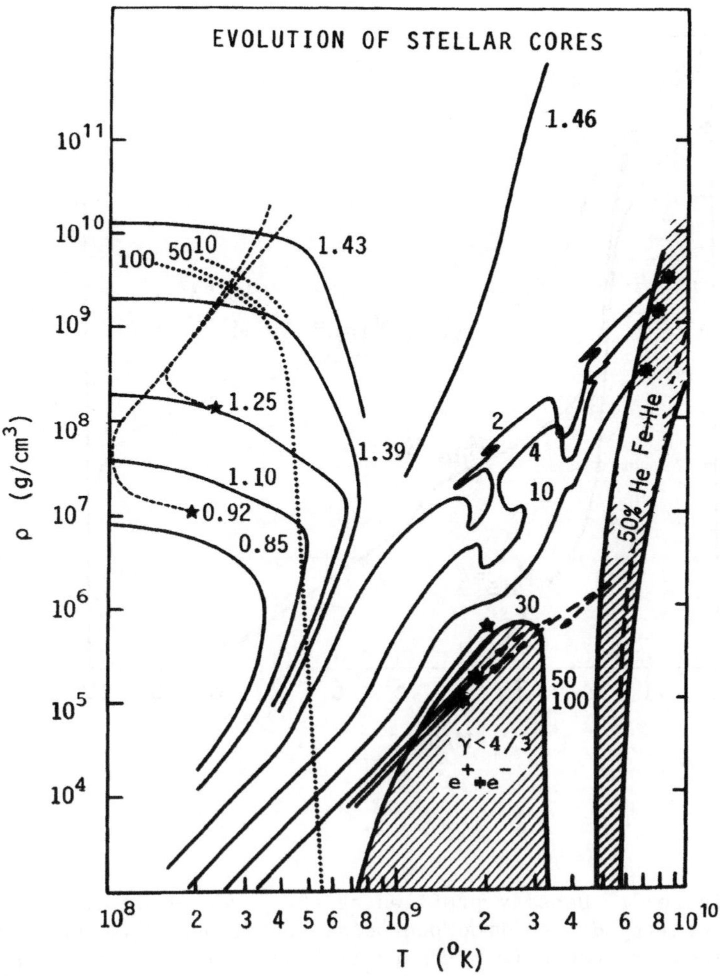

Figure 1. Evolutionary tracks of ^{16}O cores (solid lines). Masses label tracks and are assumed to be fixed. Neutrino losses are included. Stars at end of tracks represent dynamical instability. Dotted line is carbon ignition line which is given for three values of carbon mass fraction: 10, 50, and 100%. Dashed lines (upper left) represent evolution of growing carbon cores. Dashed lines (lower right) describe dynamical implosion-explosion.

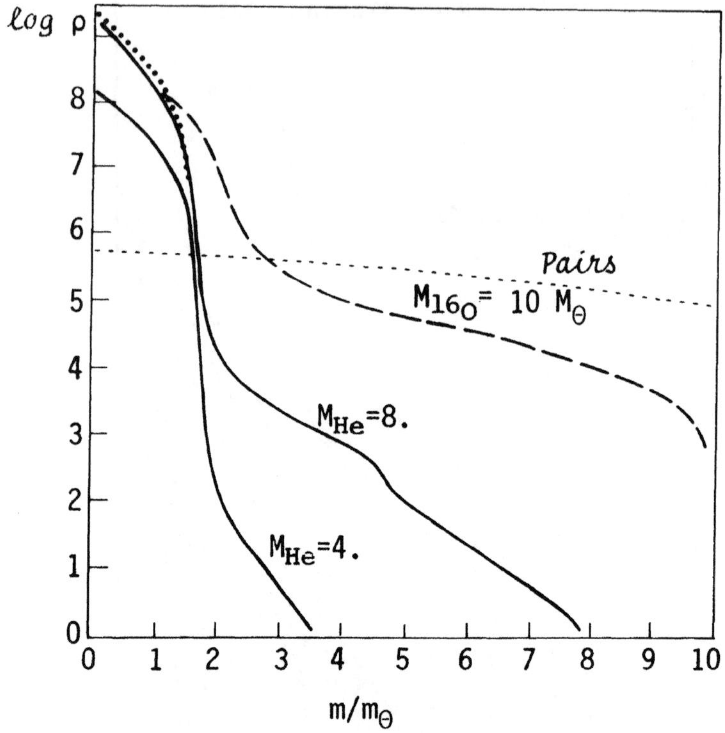

Figure 2. Density profiles at the onset of dynamical instability due to iron photodisintegration. The two solid lines are taken from Arnett 1975 and belong to the final iron cores of a 8 and 4 $M_\odot$ (initially) helium star. The dashed line is taken from Barkat 1969 and belongs to a 10 $M_\odot$ oxygen star. Dots belong to Paczynski's CDS model and the dash-dotted line is a typical pair-formation PSN.

§3. STAR-CORE RELATIONSHIP

In order to explore the question of whether fixed mass cores are indeed justifiable representations of real stellar cores, and if so which star corresponds to which core, it is of course best to carry out full evolutionary calculation. It is however well known that these are rather very difficult to perform. Again, convection and burning shells are the main pitfalls. A large effort has been recently invested by several authors to push forward such calculations, mainly for massive $M \gtrsim 10M_\odot$ stars. Evolutionary calculations through Helium burning are rather plenty. Stothers and Chin 1975, 1976 give impressive lists of references to such work whose main goal is to establish theoretically expected statistics of red vs. blue supergiants in order to compare with observation. The influence of important parameters: such as initial x, y, z values on one hand and of the different criteria for onset of convective instability, as well as various methods of treating semi-convection are being sorted out. As will be explained shortly, for our purposes it is more the variations caused in the size and composition of helium and (even more so important) Carbon-Oxygen core that are important. In this regard Stothers and Chin 1973 show that variations just due to initial x,y,z values are not trivial. Lamb et al 1976 give an interesting comparison table of works through carbon core burning. Even though this list is not quite complete it already shows how diverse results can be. Indeed it is too difficult at present to make more than an intelligent guess as to the true influence of any particular parameter because of the divergence of methods and physical input between authors and because of the fact that no one has carried out a truly systematic survey. The answer to the question of association between stellar mass and stellar core's mass can thus be at present only approximate (see for example Arnett 1973). On the other hand there can be little doubt that unless drastic changes (such as core-envelope mixing) are established, any C-O core which is massive "enough" does evolve independently of its envelope. To qualify as massive "enough" a core must first be more massive than a critical value (M_{crit}) which is close to the Chandrasekher mass ($\sim 1.44M_\odot$). This in itself may however not be sufficient. An additional key question is whether the burning shell (He) which sits just above the core is associated with a convective zone, (or else it is radiative and usually quite narrow). If indeed this is the case then it must almost always be true that the shell cannot advance before fuel consumption and this does not happen before the core evolves all the way to a stage of dynamic instability. On the other hand, radiative shells do manage to advance, and thus the core's mass is increased, before the core becomes unstable. The significance of such increase is smaler for more massive cores -- whose lifetime is much shorter. It is important to note that the same statements apply to later stages -- (i.e. carbon exhausted cores). While in the case of C-O cores there is some doubt whether for cores less massive than $\sim 3M_\odot$ the He shell is convective, there is apparently common agreement that for the more massive cores and for the later stages, in any case, this is indeed always so. It thus becomes evident that the size of the next independent core is determined by the convective shell whose base is below M_{crit} and whose top is above it. In a recent study (Reiss and Barkat 1976) it is shown that the size of a carbon exhausted core may not be a monotonic function of the C-O core's mass, and may also show a large peak. On the basis of the former considerations one can readily understand this behavior. Similarly it is shown how (and why) this size is rather sensitive to the C mass fraction (following He burning).

While there is no à priori good reason for real convergence of evolutionary tracks in the present mass range (in contrast to some of the less massive stars — see later), final (Fe-Ni) cores do look here rather similar in some respects. Since one does not know (at least yet) what are the important structural features for the dynamic stages of collapse, one cannot have a solid opinion on whether or not these final cores are indeed alike or whether the structural and/or compositional differences which occur are significant. All that can be said at this point is that their mass is within a narrow range — $1.2 \lesssim M_c/M_\odot \lesssim 2.5$ — and their density profiles are quite close to each other. Typical density profiles (cores are nearly isothermal with $T \sim 5\times10^9\,°k$) are given in Figure 2, which is taken from Arnett 1975 and Barkat 1969 (unpublished). A similar statement should be made with regard to the envelopes (Si-O-C-He-H) which are extended and relatively dilute but may have decisive importance for light curve and nucleosynthesis considerations. It is quite clear from Figure 2 that a wide range of densities exists (for different core masses), but it may be too early to speculate on its real meaning.

Our discussion has been mostly devoted to the massive stars. Typical PSN models for the very massive cores (pair formation case) are given in Figure 3 (taken from Barkat 1969, see also Arnett 1972b). A reexamination of the dynamics of these cases may be in order because of the fact that presence of carbon was not taken into account (by Barkat et al 1967, Fraley 1968). It is important to note that from the point of view of nucleosynthesis theory (explosive carbon, oxygen and silicon burning) the time dependence of the density and temperature profiles within the dynamic event are highly promising (Truran-private communication). Another unexplored aspect of this range is the further evolution of massive remnants which are left following explosion in the lower marginal region of this range.

The evolution of the less massive stars ($M \lesssim 10M_\odot$) has been less vigorously pursued in recent years (see Puczynski 1973). From the point of view of PSN there is certainly a lower mass limit to this range ($M_c \gtrsim M_{ch} \sim 1.44M_\odot$). This lower bound has to do with mass loss from the envelope and we shall return to it later. Within this range we have the well-known case of carbon detonation supernova (CDS). The existence of this PSN is mainly the result of the realization of the following conditions: 1. Formation of a C-O core whose mass is smaller than and not too close to the minimal ignition mass, at the stage of neutrino domination. 2. A slow growth of this core towards M_{ch} from below due to the advance of a thin He burning shell. These conditions combine to cause convergence of evolutionary tracks (which belong to different masses)

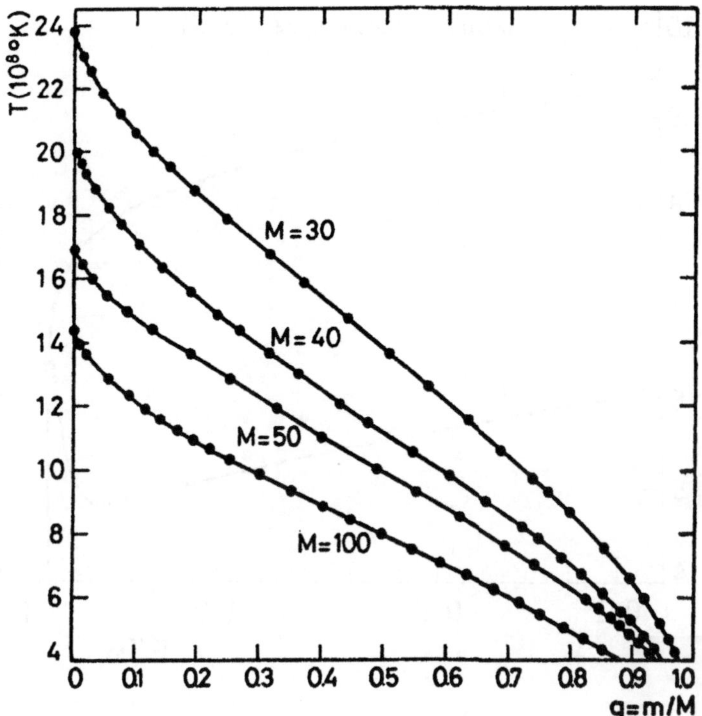

Figure 3. Temperature and density profiles of pair-formation PSN. Lines are labelled by the parent oxygen star mass. Note that the mass coordinate is mass fraction.

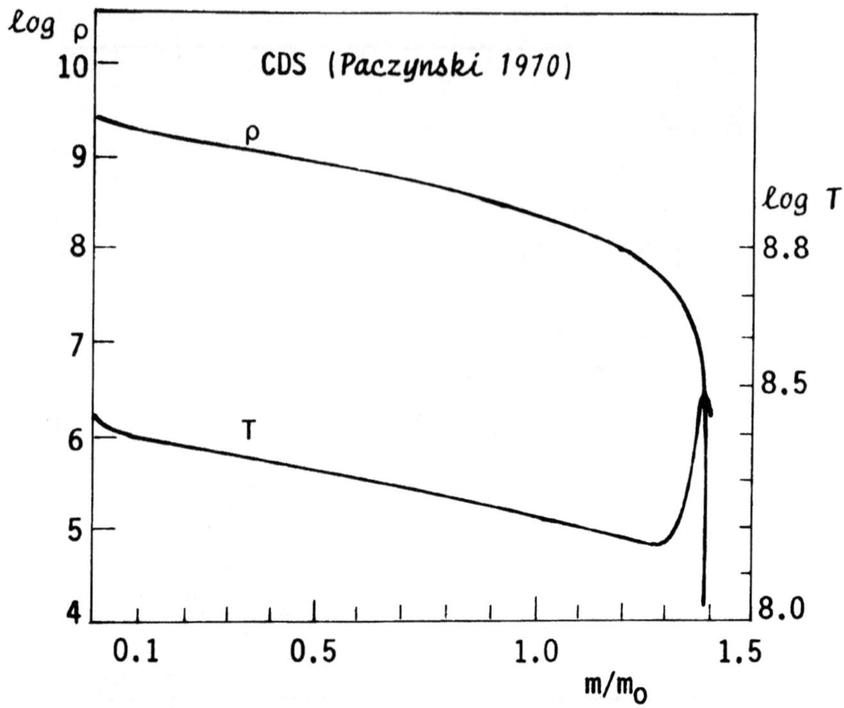

Figure 4. Density (upper line) and temperature (lower line) profiles of CDS, taken from Paczynski 1970.

and lead the growing cores to the point of explosive fuel ignition (Arnett 1969, Paczynski 1970, Barkat 1971).

While it is clear that for a mass that is low enough ($\lesssim 6M_\odot$) the first condition shall be satisfied, it is also clear that only a reduction of the He core's mass (due to either core-envelope mixing or mass loss) can be responsible for its fulfillment in the complementary higher range ($6 \lesssim M/M_\odot \lesssim 10$). This leads us naturally to the very important subject of core's mass reduction.

§4. CORE MASS REDUCTION

a. Mixing

Penetration of an outer convective zone into outer layers of He (and even C-O) cores has been encountered by few authors (Weigert 1966 for $M=5M_\odot$, Paczynski 1970 for $M=7M_\odot$). Sugimoto and Nomoto 1973 (and earlier work of Sugimoto) have analyzed this question using a special technique of fitting envelopes to cores. According to this work a decisive factor is the ratio of mixing length to pressure scale height used in the envelope. The uncertainty regarding this factor seems to leave the question open. We would also expect that variations in core structure may also be important. Improvements in this problem are certainly highly desirable.

b. Mass Loss

Mass loss from the envelopes of giants has been an important branch of astrophysics for a long time — both observationally and theoretically. It is of course beyond us to review the current state of affairs in this field. We only note that important new results (Fussi-Pecci and Renzini 1975a, Reimers 1975) may have profound effect on our present subject (beyond those suggested for instance by Woolf 1973). In particular it may set up a severe lower limit to our range (see Scalo 1976, Fussi-Pecci and Renzini 1975b and Mengle 1976). In fact it may even be possible that the lower bound exceeds the upper bound such that the whole discussion is purely academic. To add to the confusion one should note that unstable double shell burning that does surely exist at the lower part of this range may also not be negligible as hitherto assumed.

§5. CARBON DETONATION PSN

If mixing and mass loss do not destroy the evolutionary sequence leading to CDS it remains valid to continue a long and interesting road followed by workers in recent years, trying to build the correct picture of SN there. Two different subregions may then exist: a) a lower region $3 \lesssim M/M_\odot \lesssim 6$ (numbers being approximate) where the convergence of evolutionary tracks is valid and where carbon survives in the center, b) the complementary region where carbon is ignited and mostly exhausted before the center manages to evolve to ignition point (oxygen) — see Barkat et al 1974. The typical PSN structure for CDS is given in figure 4 (taken from Paczynski 1971). Here we define the PSN as the configuration at fuel ignition. The stage of evolution which should include convective Urca process would occur shortly. We shall not discuss the evolution beyond this point because we do not feel that it has yet been really satisfactorily treated -- see recent work by Lazareff 1975 and Regev and Shaviv 1976.

REFERENCES

Arnett, W.D., 1969 Ap. Space Sci. 5:180
1972a. ApJ. 176:681
1972b. ApJ. 176:699
1973. IAU Symp. No. 66, Warsaw, Poland
1974a. ApJ. 193:169
1974b. ApJ. 194:373
1975a. ApJ. 195:727

Barkat, Z., 1969 PhD Thesis, Hebrew University, Jerusalem, Israel
1971. ApJ. 163:433
1975. Ann Rev. Astron. Ap 13:45

Barkat, A., Sack, N., Rakavy, G., 1967. Phys Rev. Lett. 18:379.

Barkat, Z., Reiss, Y., Rakavy, G., 1974. ApJ. Lett. 193:L21.

Boozer, A.H., Joss, P.C., Salpeter, E.E. 1973. ApJ. 181:393.

Castelani, V., Giannone, P., Renzini, A., 1971. Ap. Space Sci. 10:34.

Endall, A.S. 1975. ApJ 197:405.

Fraley, G., 1968. Ap. Space Sci. 2:96.

Fussi-Pecci, F., Renzini, A., 1975a. Proc. 19th Liege Colloquium.
1975b. Astron. Astrophys. 39:413.

Ikeutchi, S., Nakazawa, K., Murai, T., Hoshi, R., Hayashi, C., 1971. Prog. Theor. Phys. 46:1713.
1972. Prog. Theor. Phys. 48:1870.

Imshennik, V.S., Nadyoshin, D.K., 1973. IAU Symp. No. 66, Warsaw, Poland.

Kovetz, A., Prialnik, D., Shaviv, G., 1975. Ap. Space Sci. 37:463.

Lamb, S.A., Iben, I., Howard, M., 1976, preprint.

Lazareff, B. 1975, Astron. Astrophys. 45:141.

Maeder, A. 1975. Proc. 19th Liege Colloquium.

Massevitch, A.G., Tutukov, A.V., 1973, IAU Symp. No. 66, Warsaw, Poland.

Mengel, J., 1976, Astron. Astrophys. 48:83.

Paczynski, B.E., 1970 Acta Astron. 20:47.
1971 Acta Astron. 21:1.
1973 IAU Symp. No. 66, Warsaw, Poland.

Rakavy, G., Shaviv, G., Zinamon, Z., 1967, ApJ. 150:131.

Regev, O., Shaviv, G., 1976. Astron. Astrophys. in press.

Reimers, D., 1975. Proc. 19th Liege Colloquium.

Reiss, Y., Barkat, Z., 1976, in preparation.

Savedoff, M.P., Van Horn, H., Vila, S.C., 1969, ApJ. 155:221.

Scalo, J.M., 1976, ApJ. 206:215.

Shaviv, G., Salpeter, E.E., 1973. ApJ. 184:191.

Stothers, R.B., Chin, C.W., 1973. ApJ.
1975. ApJ.
1976. ApJ.

Sugimoto, D., Nomoto, K., 1973. IAU Symp. No. 66, Warsaw, Poland.

Vila, S.C., 1966. ApJ. 146:437.
1967. ApJ. 149:613.

Weigert, A., 1966, Z. Ap. 64:395.

Woolf, N.J., 1973. IAU Symp. No. 66, Warsaw, Poland.

CMa R1: A CASE OF SUPERNOVA-INDUCED STAR FORMATION?

William Herbst
Carnegie Institution of Washington
Department of Terrestrial Magnetism

The idea that Supernovae might sometimes cause star formation is an old one. It was suggested as early as 1953 by Öpik. However, strong observational evidence in favor of this has, to my knowledge, never been presented. What one would like to find is a young, star formation region at the edge of a SNR of comparable age. In this talk I would like to present some evidence which is suggestive of such a situation.

The region of interest is known as CMa R1 and located at $\alpha=7^h00^m$, $\delta=-12°$, $\ell=225°$, $b=-2°$. It is an R-association; that is, an association of stars illuminating reflection nebulae. Its distance is known by its association with CMa OB1 which Claria has found to be 1150 pc from the sun.

On the Palomar Sky Survey prints of the region, one sees a dense, arced dust cloud extending over more than one degree, and many embedded, early-type stars. The gas is ionized along the inner edge of the arc. Many of the embedded stars have emission lines and infrared excesses, suggesting that they are extremely young objects. One in particular, should be mentioned-Z CMa-one of the classical Herbig-emission stars which are believed to be pre-main-sequence early-type stars with characteristic ages around 10^5 years.

A composite of adjacent POSS red prints shows that this arc of nebulosity in which the CMa R1 stars are situated is part of a large scale, ring-like distribution of nebulosity with a diameter of 3° (60 pc). There is no centrally located luminous star, nor are there any pulsars or x-ray sources known in that direction. However, a radio map by Day and co-workers at 2700 MHz shows a ring of sources, coincident with the optical ring, and extending along a portion of the arc where the optical nebulosity is faint or absent. Unfortunately, no spectral index information is yet available for these sources.

A natural explanation for these features would seem to be that a SN event occurred near the center of the luminous ring, pushing the interstellar material into a dense shell in which the CMa R1 stars have recently formed. It can be shown that the time scale for the formation

David N. Schramm (ed.), Supernovae. 143-144.

of a shell with diameter 60 pc is compatible with the likely age of the CMa R1 stars. Chevalier has modelled the evolution of a SNR with outburst energy E_o expanding into an initially uniform interstellar medium with density N_o. He finds that the time (t_{SNR}) necessary for the dense shell to expand to any given radius is, roughly

$$t_{SNR} \propto \left(\frac{n_o}{E_o}\right)^{0.7}$$

If we make reasonable choices for the parameters ($E_o = 3 \times 10^{50}$ ergs and $n_o = 1$ cm^{-3}), we find

$$t_{SNR} = 3 \times 10^5 \text{ years}$$

in good agreement with the likely age of the stars. This agreement shows that SN induced star formation is, in this respect, a plausible suggestion.

What is needed now is some confirmation that a SN event has actually occurred in this region. Observations of expanding motions of the gas or stars might provide this. In particular, a proper motion study of the R association members would be very valuable. It also might be expected that the region should be a source of soft x-rays. Finally, a careful study of the ionization structure and the ionization energy requirements for the emission nebulosity might produce compelling evidence in favor of the SN hypothesis.

SUPERNOVA NUCLEOSYNTHESIS

James W. Truran
Department of Astronomy
University of Illinois

1. INTRODUCTION

Theoretical studies strongly suggest that supernova environments represent the likely site of the synthesis of the bulk of the heavy elements ($A \geqslant 20$) observed in nature. From the point of view of nucleosynthesis, one has the advantage that constructive nuclear transformations proceed in the shock-induced ejection of both core and envelope material, ensuring that the resulting abundance distributions will not be distorted by subsequent evolution. Hydrodynamic studies of possible supernova mechanisms predict promising conditions both for the synthesis of elements through the vicinity of iron by charged-particle reactions (Arnett et al., 1971) and for the r-process neutron-capture synthesis of heavier nuclei (Arnett and Truran, 1970). Unfortunately, hydrodynamic studies have yet to provide a thoroughly consistent supernova model (see the review by K. Sato in these proceedings). The well studied carbon-detonation (Arnett, 1969a) and neutrino-transport (Colgate and White, 1966) mechanisms both encounter severe problems which have not yet been resolved. Dynamic events triggered by the creation of electron-position pairs (Barkat et al., 1967; Fraley, 1968) require more massive carbon-oxygen cores than are found to arise in the course of evolution of normal stars.

Given these uncertainties, I will concentrate in this review on those results of nucleosynthesis studies which I consider to be well established independent of the details of hydrodynamic models. Some speculations concerning the consequences of thermonuclear processing of calculated 'pre-supernova' stellar configurations are also presented.

2. EXPLOSIVE NUCLEOSYNTHESIS

2.1 Explosive charged-particle nucleosynthesis

Calculations of carbon, oxygen, and silicon burning proceeding at

David N. Schramm (ed.), Supernovae. 145-158. *All Rights Reserved.*

high temperatures on dynamic timescales have proved successful in their predictions of detailed abundance features (Arnett 1973a; Truran 1973). In these studies, matter of initial composition consistent with post helium-burning conditions in stellar cores is subjected to temperatures and densities compatible with its ejection in a supernova event. For an initial composition consisting of equal parts by mass ^{12}C and ^{16}O, with a small admixture of the neutron-rich isotope ^{22}Ne (formed from ^{14}N during helium burning), the following results are obtained (Truran and Arnett, 1971):

Explosive "carbon" burning ($T \sim 2 \times 10^9$ K, $\rho \sim 10^5$ g/cc) forms many of the observed abundance features in the mass range $20 \leq A \leqslant 30$ (Arnett, 1969b).

Explosive "oxygen" burning ($T \sim 3.6 \times 10^9$ K, $\rho \sim 5 \times 10^5$ g/cc) will reproduce most of the observed features (Figure 1) in the mass range $28 \leq A \leqslant 44$ (Truran and Arnett, 1970; Woosley et al., 1973).

Explosive "silicon" burning ($T \sim 5 \times 10^9$ K, $\rho \sim 10^6$ g/cc) produces nuclei in the iron-peak region (Figure 2) $48 \leqslant A \leqslant 60$ in solar proportions (Truran et al., 1967; Truran and Arnett 1971).

The names of these various burning processes are carryovers from the corresponding nuclear burning stages of stellar evolution. The feature which distinguishes these three explosive processes is, rather, the peak shock temperature to which the matter is subjected. It is likely that such burning proceeds simultaneously, but in different mass zones, in the supernova of a massive star. If this occurs, the ejected material will consist of a mix of the three distributions and may reproduce, in a single event, the major abundance features in the mass range $20 \leq A \leqslant 60$ (Arnett and Schramm, 1973).

The strongest argument for the origin of nuclei from neon to nickel in such explosive phases arises from the manner in which they reproduce accurately determined terrestrial isotopic abundances. The stable isotopes of chlorine, for example, are formed in these explosions as ^{35}Cl and ^{37}Ar, yet following ^{37}Ar decay the terrestrial isotopic ratio $^{35}Cl/^{37}Cl$ is reproduced to better than 50 percent. This same phenomenon is exhibited in many other cases: the potassium isotopes ^{39}K and ^{41}K (formed as ^{41}Ca); the chromium isotopes ^{50}Cr, ^{52}Cr and ^{53}Cr (^{52}Cr and ^{53}Cr formed in situ as ^{52}Fe and ^{53}Fe); the iron isotopes ^{54}Fe, ^{56}Fe and ^{57}Fe (^{56}Fe and ^{57}Fe produced as ^{56}Ni and ^{57}Ni); and the nickel isotopes ^{58}Ni, ^{60}Ni, ^{61}Ni and ^{62}Ni (the last three formed as ^{60}Zn, ^{61}Zn and ^{62}Zn). Furthermore, the relative abundances of ^{51}V (formed as ^{51}Mn), ^{55}Mn (formed as ^{55}Co), and ^{59}Co (formed as ^{59}Cu) are reproduced very well in these calculations. Having found nature to be deceptive but never malicious, I interpret this detailed agreement of theory and observation as strong evidence in support of the operation of these explosive nucleosynthesis mechanisms.

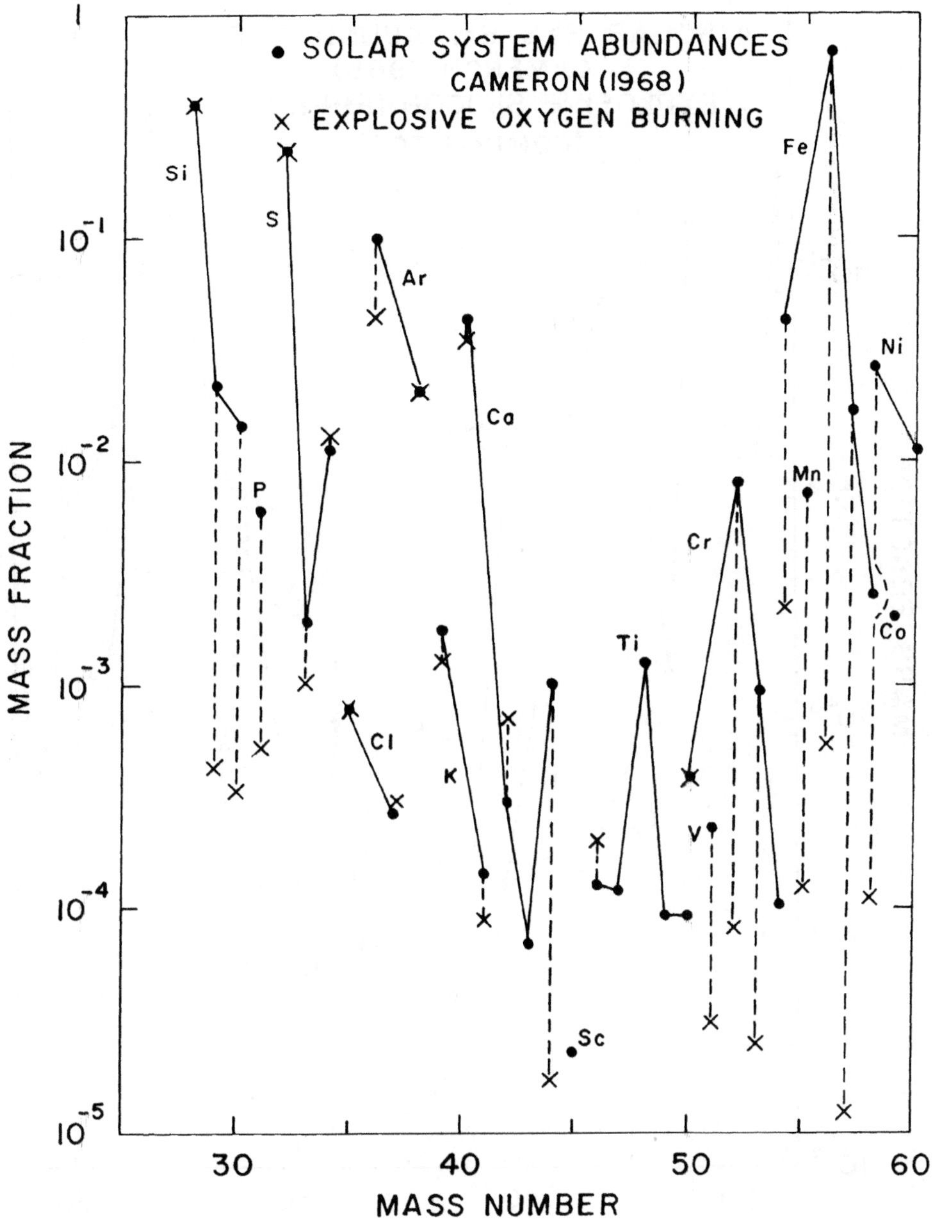

Figure 1. The abundance pattern achieved in explosive oxygen burning compared to solar abundances. (Normalization at ^{28}Si.)

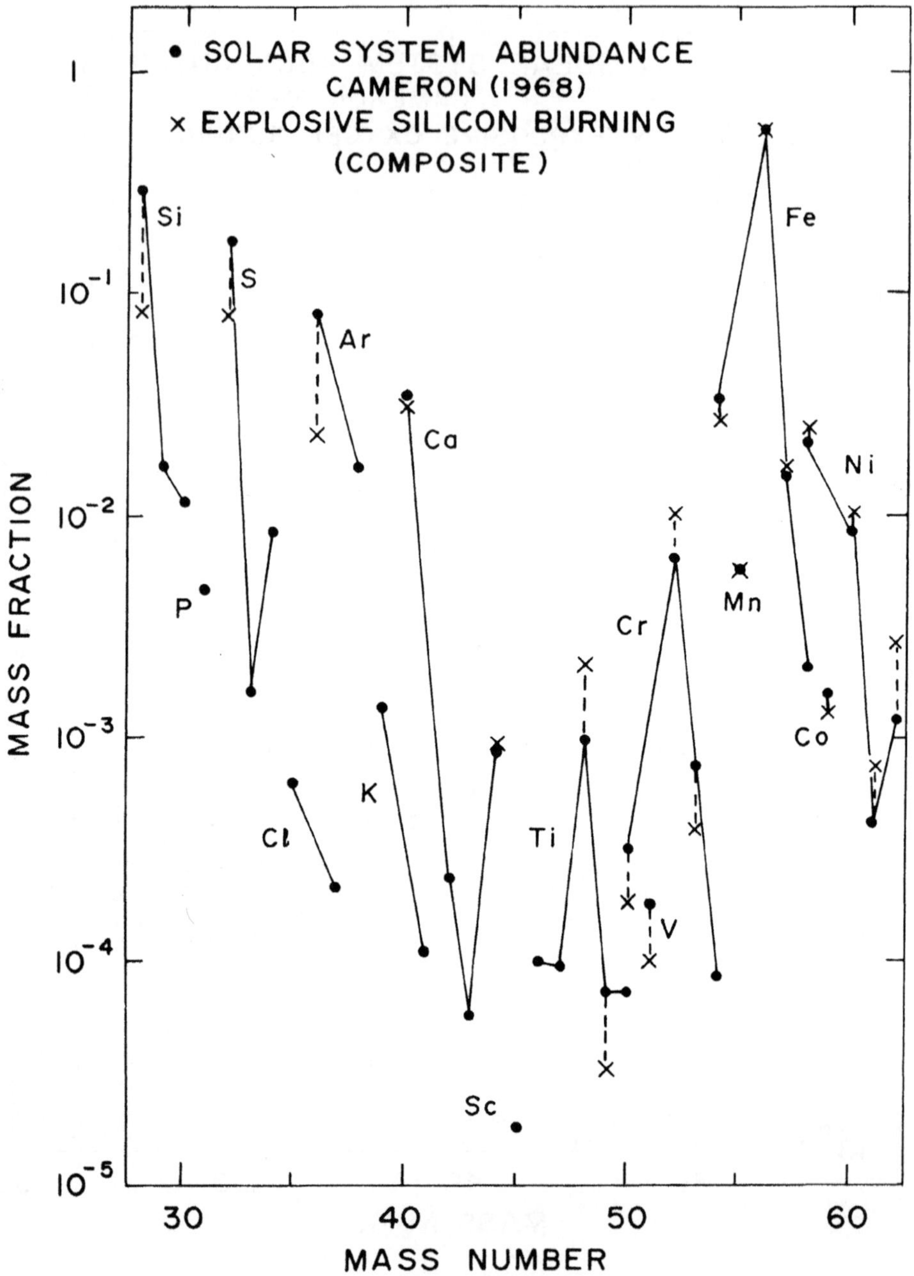

Figure 2. The abundance pattern achieved in explosive silicon burning compared to solar abundances. (Normalization at ^{56}Fe.)

2.2 Explosive neutron-capture nucleosynthesis

Supernovae also provide the most promising environment for the production of heavy nuclei by r-process neutron-capture synthesis. Matter arising from the vicinity of the mass cut in supernovae which give rise to neutron star remnants can be expected to be highly neutronized. The models of Colgate and White (1966) and of Arnett (1967) predict total ratios of neutrons to protons reaching $N/Z \sim 2\text{-}8$, established by weak interaction processes at temperatures $T > 10^{10}$ K and densities $\rho > 10^{10}$ g/cc. It has been demonstrated that the subsequent thermonuclear evolution of this matter through expansion and cooling will give rise to the formation of heavy nuclei ($A > 60$) by rapid neutron captures. In spite of the problems associated with these specific hydrodynamic studies, it is instructive to review the nuclear behavior: ejection of such neutronized matter is predicted by other hydrodynamic studies as well (Le Blanc and Wilson, 1970; Meier et al., 1976).

For temperatures above 10 billion degrees, photonuclear reactions assure that the matter will be composed largely of free neutrons and protons. Cooling gradually allows thermonuclear buildup first of helium and then of heavier elements. When the temperature has fallen to $T \sim 3\text{-}5 \times 10^9$ K, nuclear statistical equilibrium distributions dominated by extremely neutron-rich nuclei are expected for the large neutron-to-proton ratios which characterize this matter (Tsuruta and Cameron, 1965). Studies of the thermonuclear transformations defining the approach to these equilibrium configurations (Truran et al., 1968; Delano and Cameron, 1971) have established that they can be achieved on the hydrodynamic expansion timescale.

Neutron-rich equilibria at temperatures of a few billion degrees, for total neutron-to-proton ratios $N/Z \sim 2\text{-}8$, are also characterized by large concentrations of free neutrons. The subsequent nuclear evolution of this expanding matter has been demonstrated in "dynamical" r-process calculations (Cameron et al., 1970; Schramm, 1973; Hillebrandt et al., 1976) to form heavy nuclei through the transuranic region. The gross features, if not every detail, of solar system r-process abundances are reasonably reproduced in such calculations. Both the positions of the r-process abundance peaks and the absence of substantial odd-even abundance variations can be generally understood. The considerable uncertainties associated with extrapolations of nuclear properties to extremely neutron-rich nuclei--in the vicinity of the neutron-drip line--assure that definitive treatments are not yet possible. I believe, nevertheless, that the dynamic ejection of neutronized matter from supernova cores provides the most likely astrophysical environment for the r-process.

2.3 Hydrogen- and helium-induced nucleosynthesis

Several further charged-particle nucleosynthesis mechanisms can operate in supernovae. Howard et al. (1971) have considered the reactions which can occur in the explosive ejection of the helium shells of supernovae. They find that the production of ^{15}N, ^{18}O, ^{19}F, and ^{21}Ne occurs under such conditions. Lazareff et al. (1977) have determined that ^{15}N and ^{17}O represent the dominant products of CNO-cycle hydrogen burning such as can occur at the base of the hydrogen envelope following shock heating to temperatures $T \gtrsim 2 \times 10^8$ K. The results of these studies, while suggestive, are far more sensitive to uncertainties in the envelope structure and shock-imposed burning conditions, and must therefore be considered more speculative.

Several recent investigations have been concerned with the production of the $\sim$ 35 heavy nuclei ($A > 60$) which cannot be formed by neutron-capture mechanisms. These "bypassed" nuclei lie on the proton-rich side of the valley of beta stability, with an abundance level typically $\sim$ 1/100 that of neighboring neutron-capture products. Their production is generally attributed to some combination of proton-capture (p-process) and photoneutron reactions involving the products of earlier neutron-capture synthesis as seeds. A consideration of the rates of appropriate proton-induced and neutron photodisintegration reactions (Truran and Cameron, 1972) reveals that temperatures $T \sim 2 \times 10^9$ K are required, if significant production of p-process nuclei is to occur on a hydrodynamic time scale. Audouze and Truran (1975) have calculated the abundances resulting for representative expansion profiles in matter of solar composition (Cameron, 1973), assumed representative of supernova envelopes. The enhancements achieved as a function of peak shock temperature are shown in the table. Comparable overproductions of most of the designated p-process nuclei are predicted for the case $T = 2 \times 10^9$ K. Unfortunately, such high temperatures are incompatible with shock heating of the low density hydrogen envelopes of supernovae: initial densities $\rho \gtrsim 10^4$ g/cc are required, as one can verify directly by consideration of the Hugoniot-Rankine shock relations. Unpublished calculations by W. M. Howard and S. A. Woosley (private communication) suggest that p-process nucleosynthesis may rather occur during phases of explosive carbon and oxygen burning in inner mass shells ejected in supernovae, where the required temperature-density conditions are expected to exist.

3. SUPERNOVA MODELS

A summary of predicted supernova core nucleosynthesis conditions is presented in Figure 3. The gross features of the abundance distributions resulting from the thermonuclear processing of the ejecta may be inferred from these representative expansion profiles.

3.1 Neutrino-transport mechanism

For the neutrino-transport models (the behavior of representative zones for the models of Colgate and White (1966) and of Arnett (1967) are indicated), the critical consideration is that the densities in the innermost ejected mass zones must be in excess of $\sim 10^{10}$ g/cc if sufficient energy is to be deposited in these regions by neutrinos leaving

TABLE

p-PROCESS ENHANCEMENTS

NUCLEUS	Ratios of Abundances to Solar-System Abundances		
	1.5×10^9 K	2×10^9 K	2.5×10^9 K
^{74}Se	44.0	46.0	47.0
^{78}Kr	270.0	340.0	390.0
^{84}Sr	270.0	280.0	270.0
^{92}Mo	45.0	32.0	24.0
^{94}Mo	($\leqslant$ 1)	($\leqslant$ 1)	(< 1)
^{96}Ru	430.0	330.0	380.0
^{98}Ru	20.0	19.0	20.0
^{102}Pd	45.0	55.0	90.0
^{106}Cd	22.0	20.0	34.0
^{108}Cd	60.0	120.0	210.0
^{112}Sn	10.0	($\leqslant$ 1)	(< 1)
^{113}In	($\leqslant$ 1)	($\leqslant$ 1)	(< 1)
^{114}Sn	13.0	15.0	170.0
^{115}Sn	39.0	18.0	(< 1)
^{120}Te	190.0	160.0	1.1
^{124}Xe	140.0	190.0	86.0
^{126}Xe	150.0	190.0	280.0
^{130}Ba	230.0	250.0	500.0
^{132}Ba	200.0	500.0	850.0
^{136}Ce	61.0	740.0	310.0
^{138}Ce	78.0	38.0	380.0
^{138}La	($\leqslant$ 1)	($\leqslant$ 1)	($\leqslant$ 1)
^{144}Sm	120.0	320.0	550.0
^{146}Sm	57.0	500.0	750.0
^{152}Gd	6.4	4.7	($\leqslant$ 1)
^{156}Dy	(< 1)	62.0	(< 1)
^{158}Dy	(< 1)	200.0	(< 1)
^{162}Er	(< 1)	320.0	(< 1)
^{164}Er	(< 1)	42.0	(< 1)
^{168}Yb	(< 1)	360.0	(< 1)
^{174}Hf	(< 1)	140.0	(< 1)
^{180}Ta	(< 1)	($\leqslant$ 1)	(< 1)
^{180}W	(< 1)	120.0	(< 1)
^{184}Os	(< 1)	82.0	(< 1)
^{190}Pt	(< 1)	91.0	(< 1)
^{196}Hg	(< 1)	90.0	(< 1)

the supernova core. At these densities and temperatures ($T \geqslant 10^{10}$ K) the matter becomes extremely neutronized ($N/Z \geqslant 2$) as a consequence of weak interactions, giving rise to neutron-rich equilibra following expansion and cooling. Subsequently, neutron-capture buildup of heavy

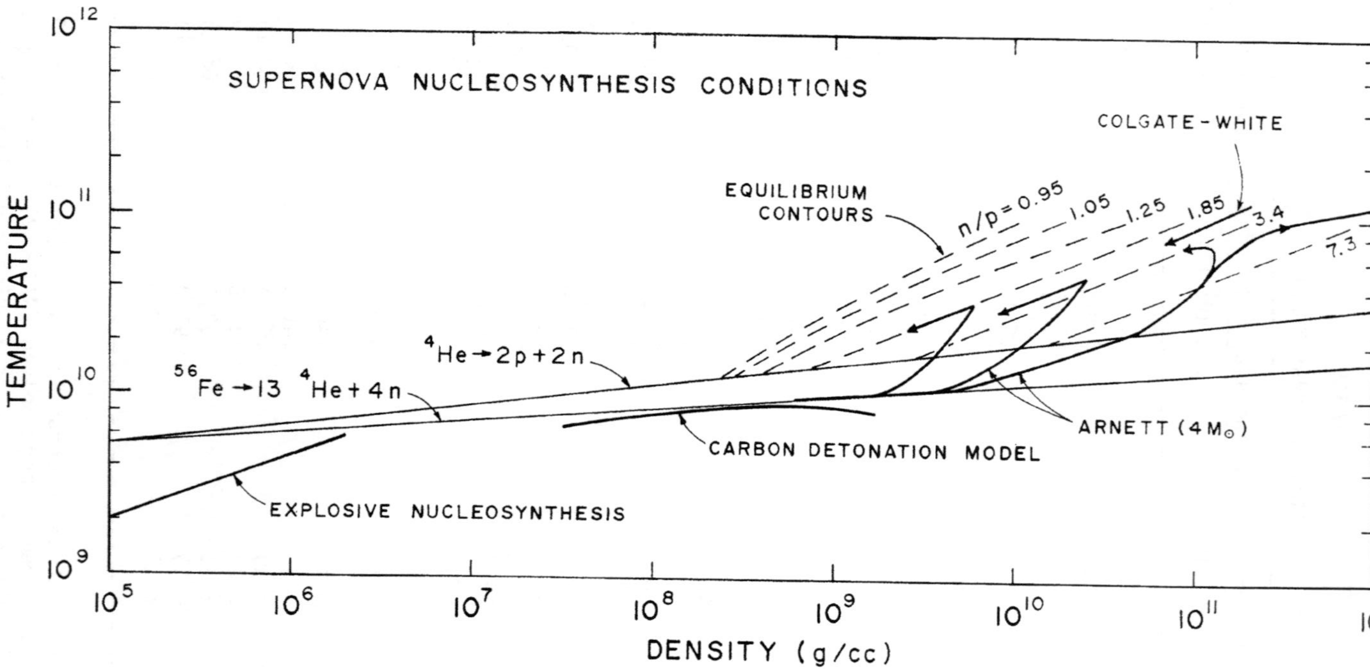

Figure 3. Survey of predicted supernova nucleosynthesis conditions for ejected core matter. Contours of constant neutron/proton ratio in weak interaction equilibrium are indicated, as are contours defining regions dominated by ^{56}Fe, ^{4}He, and neutrons and protons, respectively, in nuclear statistical equilibrium.

nuclei occurs. Independent of questions concerning the viability of this supernova mechanism, the associated r-process nucleosynthesis studies may well be relevant. Some supernova mechanism clearly gives rise to neutron star remnants (pulsars), and any ejecta arising from the vicinity of the mass cut will quite likely be highly neutronized. The magnetic-jet mechanism described by LeBlanc and Wilson (1970; see also Meier et al., 1976) predicts comparable nucleosynthesis conditions in expanding neutronized matter.

3.2 Carbon detonation mechanism

The carbon detonation supernova mechanism (the indicated curve represents the locus of peak temperatures and densities achieved in the ejecta (Arnett, 1969a)) is triggered by carbon ignition in an electron-degenerate core configuration, the central density ranging up to $\sim 10^{10}$ g/cc. As a consequence of detonation, temperatures in the range $T \sim 4\text{-}10 \times 10^9$ K are realized throughout the core and total disruption of the core ($\sim 1.4\ M_\odot$) is predicted to occur. Nucleosynthesis studies (Arnett, Truran, and Woosley, 1971; Bruenn, 1972) have demonstrated that the ejecta will be composed almost exclusively of iron-peak nuclei. The difficulties with this picture are apparent: since the predicted progenitors of these detonation events (stars of main sequence masses $\sim 3\text{-}8\ M_\odot$) are relatively numerous, implying they should represent one of the more common supernova types, neither the absence of a remnant nor the iron-peak dominated ejecta are particularly attractive features. Proposed alternatives to total disruption vary widely (Buchler et al., 1974), from considerations of convectively driven URCA neutrino losses (Paczynski, 1972) to suggestions that red-giant mass loss may be sufficient to ensure that the pre-detonation core configurations cannot be realized in the normal course of stellar evolution.

3.3 Massive star supernovae

In contrast to the extreme conditions demanded by the neutrino-transport and detonation mechanisms, promising conditions inferred from studies of explosive carbon-, oxygen- and silicon-burning nucleosynthesis define a curve at significantly lower densities (Figure 3). Such low density configurations are typical of massive stars during their late stages of evolution. Unpublished calculations by Z. Barkat and this writer indicate that the range of temperature-density conditions resulting from the explosion of massive carbon-oxygen cores, wherein dynamic instability is triggered by the creation of electron-positron pairs (Fraley 1968; Barket, Rackavy and Sack 1967), is generally consistent with the formation of nuclei from neon to iron by charged particle thermonuclear reactions. It remains to be demonstrated that such presupernova configurations can result in the normal course of stellar evolution.

Arnett (1973b) has calculated the evolution of helium burning cores of masses $M_\alpha/M_\odot = 2,4,8,16$, and 32 (corresponding by his estimates to stars of total masses $\sim 10\text{-}70\ M_\odot$). Arnett and Schramm (1973) have

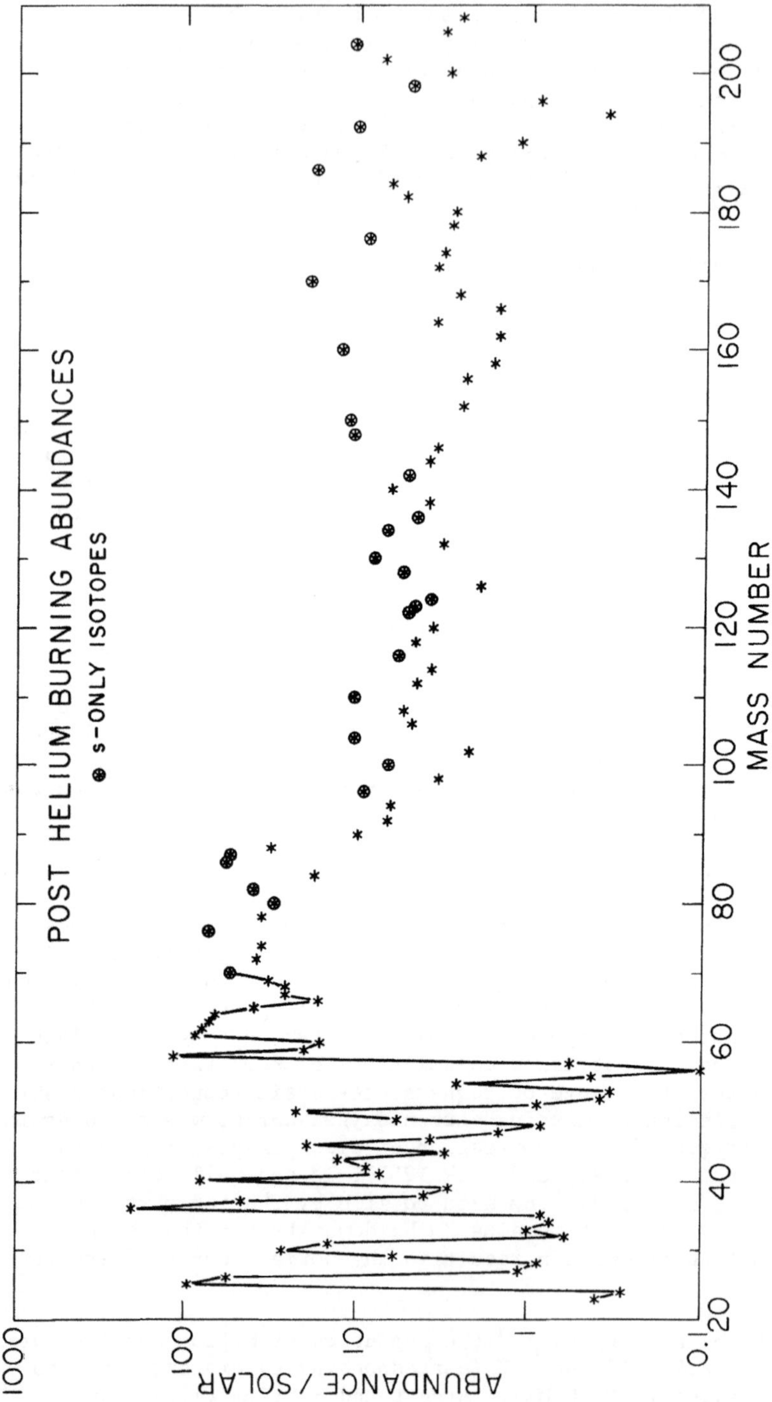

Figure 4. Heavy element abundances (relative to solar) following helium burning in the core of a 25 $M_\odot$ star. The neutron exposure was provided by the $^{22}Ne(\alpha,n)^{25}Mg$ reaction.

estimated that the mass average of all matter predicted to be ejected following thermonuclear disruption, assuming a Salpeter initial mass function, gives solar system proportions of ^{12}C, ^{16}O, and nuclei in the range ^{20}Ne to ^{60}Ni by means of explosive nucleosynthesis. This predicted agreement is encouraging, but it must be remembered that hydrodynamic studies have thus far failed to demonstrate that stars over this entire mass range will indeed give rise to supernova. In the absence of detailed hydrodynamic predictions of the post-shock temperatures imparted to this matter, any abundance predictions must be considered uncertain.

4. IMPLICATIONS OF ^{26}Mg ANOMALIES IN METEORITES

Lee et al. (1976) have recently determined that an excess of ^{26}Mg exists in inclusions in the Allende carbonaceous meteorite which are rich in aluminum and depleted in magnesium. They interpret this as evidence that ^{26}Al existed in primitive solar system material. This result may have far-reaching implications for supernova nucleosynthesis and cosmochronology. ^{26}Al is an extinct radio nuclide having a half-life of only 7.2×10^5 years. It can be formed in supernovae as a consequence of explosive carbon burning, in ratios to stable ^{27}Al which, based upon unpublished estimates by this writer, lie in the range $^{26}Al/^{27}Al \sim 5 \times 10^{-4}$ to 2×10^{-3}. Lee et al. (1976) have estimated the initial solar system ratio to be $^{26}Al/^{27}Al = 6 \times 10^{-5}$. Thus, if their detected ^{26}Al represents a product of supernova nucleosynthesis, it must have been formed in a supernova event which cannot have preceded the formation of the solar system by more than a few million years.

Cameron and Truran (1976) have considered the implications of this 'last minute' supernova contamination for other critical abundances as well. Assuming a massive star supernova (Arnett, 1973b), necessary to provide the ^{26}Al, they estimate that observed oxygen and neon anomalies may also be attributable to this supernova event. Furthermore, importan concentrations of ^{40}K and of the short-lived radionuclei ^{129}I (1.7×10^7 years) and ^{244}Pu (8.3×10^7 years) should also be present in the supernova ejecta. Production of these nuclei results from several neutron-capture episodes associated with the thermonuclear history of the ejected matter.

During the phase of core helium burning for stars above $\sim 15\ M_\odot$, the $^{22}Ne(\alpha,n)^{25}Mg$ reaction can provide a sufficient neutron flux for s-processing. Lamb et al. (1977) have calculated the resulting nucleosynthesis for a star of $25\ M_\odot$. An initial solar system distribution of abundances is found to be modified as illustrated in Figure 4. Here, the ratios of calculated to solar abundances are plotted as a function of mass number in the range $20 \leqslant A \leq 208$. In the mass region $A > 70$, isotopes which can be formed only by s-process neutron capture synthesis are designated *. The substantial overabundance realized at mass $A = 40$ is that of ^{40}K. Its production is a natural consequence of such burning phases and is expected to occur over a broad range of stellar masses. Cameron and Truran (1976) estimate that the further exposure of this matter to neutron fluxes provided by explosive burning in the carbon and helium shells in the ejection of the envelope in a supernova event will produce substantial concentrations of ^{129}I and ^{244}Pu. When account is taken of dilution associated with mixing of the supernova debris into the primitive solar nebula, these concentrations are found to be sufficient to explain the estimated primordial abundances of these radionuclei. If these estimates prove, upon further study, to be correct, it must follow that ^{129}I and ^{244}Pu dating cannot be relevant to considerations of galactic cosmochronology (Schramm, 1974).

5. CONCLUSIONS

Definitive statements regarding details of supernova nucleosynthesis must certainly wait upon refined studies of stellar evolution and supernova hydrodynamics. The procedures required to trace the thermonuclear history of ejected mass zones are well established, and their reliability can only improve with time as further nuclear experiments provide critical reaction rates. The general conclusion that supernovae provide the astrophysical environments in which the bulk of the heavy elements observed in nature are formed by thermonuclear mechanisms is unlikely to require modification.

ACKNOWLEDGEMENTS

This research was supported in part by the National Science Foundation, Grant AST73-05117, at the University of Illinois.

REFERENCES

Arnett, W. D.: 1967, Can. J. Phys. **45**, 1621.

Arnett, W. D.: 1969a, Astrophys. Space Sci. **5**, 180.

Arnett, W. D.: 1969b, Astrophys. J. **157**, 1369.

Arnett, W. D.: 1973a, Ann. Rev. Astron. Astrophys. **11**, 73.

Arnett, W. D.: 1973b, in D. N. Schramm and W. D. Arnett (eds.), Explosive Nucleosynthesis, Univ. of Texas Press, Austin, p. 236.

Arnett, W. D. and Schramm, D. D.: 1973, Astrophys. J. Letters **184**, L47.

Arnett, W. D. and Truran, J. W.: 1970, Astrophys. J. **160**, 959.

Arnett, W. D., Truran, J. W., and Woosley, S. E.: 1971, Astrophys. J. **165**, 87.

Audouze, J. and Truran, J. W.: 1975, Astrophys. J. **202**, 204.

Barkat, Z., Rakavy, G., and Sack, N.: 1967, Phys. Rev. Letters **18**, 379.

Bruenn, S. W.: 1972, Astrophys. J. Suppl. **24**, 283.

Buchler, J. R., Mazurek, T. J., and Truran, J. W.: 1974, Comments Astrophys. Space Phys. **7**, 45.

Cameron, A.G.W.: 1973, Space Sci. Rev. **15**, 121.

Cameron, A.G.W., Delano, M. D., and Truran, J. W.: 1970, in International Conference on the Properties of Nuclei far from the Region of Beta-Stability, CERN, p. 735.

Cameron, A.G.W. and Truran, J. W.: 1976, Icarus (in press).

Colgate, S. A. and White, R. H.: 1966, Astrophys. J. **143**, 626.

Delano, M. D. and Cameron, A.G.W.: 1971, Astrophys. Space Sci. **10**, 203.

Fraley, G. S.: 1968, Astrophys. Space Sci. **2**, 96.

Hillebrandt, W., Takahashi, K., and Kodama, T.: 1976, Astron. Astrophys. (in press).

Howard, W. M., Arnett, W. D., and Clayton, D. D.: 1971, Astrophys. J. **165**, 495.

Lamb, S. A., Howard, W. M., Truran, J. W., and Iben, I., Jr.: 1977, "Nucleosynthesis in the Hydrogen-Exhausted Core of a 25 $M_{\odot}$ Star" (in preparation).

Lazareff, B., Audouze, J., Starrfield, S. and Truran, J. W.: 1977, "Hot CNO-Ne Cycle Hydrogen Burning II. Explosive Hydrogen Burning" (in preparation).

LeBlanc, J. M. and Wilson, J. R.: 1970, Astrophys. J. **161**, 541.

Lee, T., Papanastassiou, D. A., and Wasserburg, G. J.: 1976, Geophys. Res. Letters **3**, 109.

Meier, D. L., Epstein, R. I., Arnett, W. D., and Schramm, D. N.: 1976, Astrophys. J. **204**, 869.

Paczynski, B.: 1972, Astrophys. Letters **11**, 53.

Schramm, D. N.: 1973, in D. N. Schramm and W. D. Arnett (eds.), Explosive Nucleosynthesis, Univ. of Texas Press, Austin, p. 84.

Schramm, D. N.: 1974, Ann. Rev. Astron. Astrophys. **12**, 383.

Truran, J. W.: 1973, Space Sci. Rev. **15**, 23.

Truran, J. W. and Arnett, W. D.: 1970, Astrophys. J. **160**, 181.

Truran, J. W. and Arnett, W. D.: 1971, Astrophys. Space Sci. **11**, 430.

Truran, J. W., Arnett, W. D., and Cameron, A.G.W.: 1967, Can. J. Phys. **45**, 2315.

Truran, J. W., Arnett, W. D., Tsuruta, S., and Cameron, A.G.W.: 1968, Astrophys. Space Sci. **1**, 129.

Truran, J. W. and Cameron, A.G.W.: 1972, Astrophys. J. **171**, 89.

Tsuruta, S. and Cameron, A.G.W.: 1965, Can. J. Phys. **43**, 2056.

Woosley, S. E., Arnett, W. D., and Clayton, D. D.: 1973, Astrophys. J. Suppl. **26**, 231.

GRAVITATIONAL COLLAPSE, WEAK INTERACTIONS, AND SUPERNOVA OUTBURSTS

V. M. Chechetkin, V. S. Imshennik, L. N. Ivanova, and
D. K. Nadyozhin
Institute of Applied Mathematics, U.S.S.R. Academy of
Sciences, Moscow, U.S.S.R.

ABSTRACT

The various mechanisms of dynamic instability onset at the last stages of stellar evolution are surveyed. The calculations of gravitational collapse of iron-oxygen stars are discussed. It is shown that there is no envelope ejection in this case, even when neutrino-nuclear coherent scattering is taken into account. The calculations of the carbon flash dynamics result in deflagrative regime of carbon burning. The carbon burning is finished up with formation of a collapsed, gravitationally bound, remnant when initial central density exceeds (3 - 5) 10^9 g cm^{-3}. The modelling of the supernova outbursts gives us an evidence that the observational data for type I supernovae are well reproduced by slow (for several days) pumping of energy into envelope.

1. INTRODUCTION

Two reasons of arising of hydrodynamic instability at final stages of stellar evolution after complete exhaustion of nuclear energy (just after formation of the iron core with mass M_{Fe}) are presently being considered:

I. For $M_{Fe} \geq 2\ M_\odot$ the instability is a result of dissociation of iron nuclei into alpha-particles and free nucleons (a quick thermodynamically equilibrium process).

II. For $M_{Ch} \approx 1.1\ M_\odot \leq M_{Fe} \leq 1.2\ M_\odot$ the onset of the instability is due to the nonequilibrium neutronization of matter, the time scale of the neutronization being of the order of $t_\beta \approx 10^5$ sec. It is much greater than hydrodynamic time scale $t_H \approx 1$ sec. The case $1.2 \leq M_{Fe} < 2\ M_\odot$ is the intermediate one. Both the dissociation of iron nuclei and the neutronization may be of importance here.

According to modern theory of stellar evolution the masses of iron cores are in the range of $1\ M_\odot \lesssim M_{Fe} \lesssim 3\ M_\odot$. It is, however, not

David N. Schramm (ed.), Supernovae. 159-182.

excluded that the more massive iron cores exist, for example, on account of mixing of stellar matter due to circulation currents. The consideration of iron cores with masses up to 10 $M_\odot$ is, therefore, our interest.

The third possibility of the dynamic instability onset is intensively being investigated in recent years. It deals with carbon burning in a degenerated stellar core within hydrodynamic time scale (Hansen and Wheeler, 1969; Arnett, 1969; Bruenn, 1972; Ivanova et al., 1974). Arnett's work (1969) has stimulated a number of subsequent investigations in this direction.

In stars with masses at main sequence ($3.5\ M_\odot < M_{ms} < 8\ M_\odot$) at the final stages of evolution there are developing degenerated carbon-oxygen cores with the approximately same mass $M_{CO} \approx 1.4\ M_\odot$ (Rose, 1969; Paczynski, 1970; Uus, 1970).

The next evolutionary stage of such stars -- the stage of carbon burning or carbon flash -- likely proceeds in hydrodynamic regime. Thus, there is the third possibility of onset of instability:

III. The carbon flash in the degenerated stellar core of the mass $M_{CO} \approx 1.4\ M_\odot$ which ultimately results in the loss of hydrostatic equilibrium.

Let us discuss in details the results of recent calculations for three kinds of dynamic instability listed above.

2. THE COLLAPSE OF IRON-OXYGEN STARS

The collapse of iron stellar core due to the dissociation of iron nuclei has a highly nonhomogeneous character. At first, a small collapsing core, which is a progenitor of a hot neutron star, is formed in stellar interiors, and then the remaining external matter is gradually accreting on the core. The two following mechanisms of ejection of stellar envelope, i.e. of supernova outburst, were suggested:

(a) the ignition of nuclear fuel remnants (oxygen) in the envelope of a collapsing star (Fowler and Hoyle, 1965), and (b) the transfer of gravitational energy from the collapsing core to the envelope by means of neutrino deposition (Colgate and White, 1966).

Ivanova et al. (1969) first fulfilled the self-consistent calculations of oxygen burning kinetics backgrounded by gravitational collapse. The weak efficiency of this process was found out. There was, nevertheless, a hope to obtain the envelope outburst, observed for supernovae, with the help of simultaneous action of deposition and oxygen burning (neutrino-thermonuclear model, Imshennik and Nadyozhin, 1974). Wilson (1971) has found that pure deposition, without nuclear burning, was ineffective for envelope ejection. The derivation of neutrino heat conductivity equations (Imshennik and Nadyozhin, 1972) has led to further progress in investigation of gravitational collapse.

Having used these equations, Nadyozhin (1975a,b) calculated the collapse of iron-oxygen stars 2 $M_\odot$ and 10 $M_\odot$. These works develop in many respects the paper by Ivanova et al. (1969). Figures 1-3 show the velocity, temperature, and density versus Lagrangian coordinate for the star of 2 $M_\odot$ at the intermediate stage of collapse just after rise of neutrino opaque stellar core. Both the mass scale and the mesh point numbers are indicated in Figure 1. The star was all divided at 151 mesh points (the 10 external mesh points, containing 4 percent of total stellar mass, are not indicated in Figures 1-3).

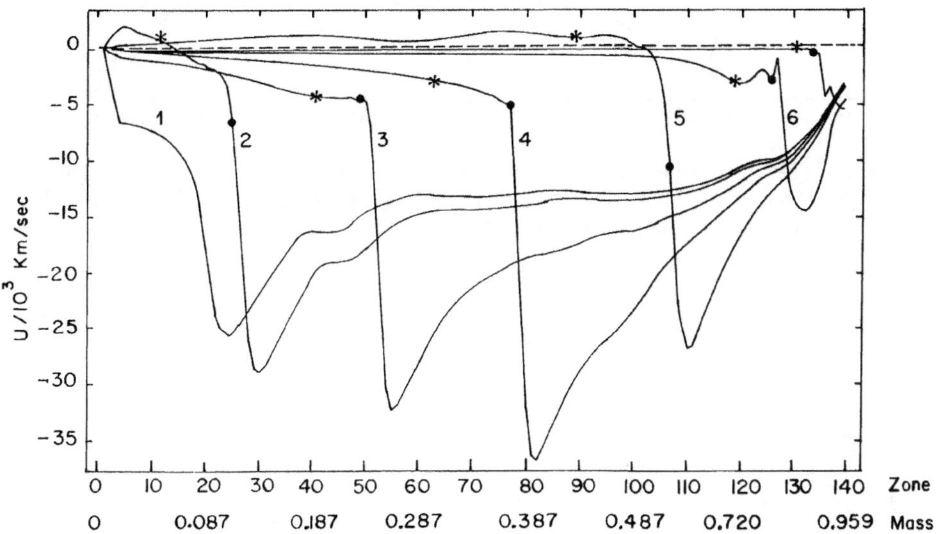

Figure 1. Velocities versus Lagrangian coordinates in collapsing iron-oxygen star of 2 $M_\odot$ for various moments of time just after rising the core which is opaque to neutrino radiation. J denotes the number of mesh points and m is the relative mass variable. Marker • is used for boundary of the opaque core, and marker * indicates the position of neutrino photosphere. The correspondence between curve number and physical time t is given by the next table:

curve number	1	2	3	4	5	6	7
$t/10^{-2}$s	0.191	0.248	0.428	0.648	1.06	3.68	12.5

The time is measured from the moment when total neutrino optical depth equals $t_{\nu\tilde{\nu}} = 0.001$.

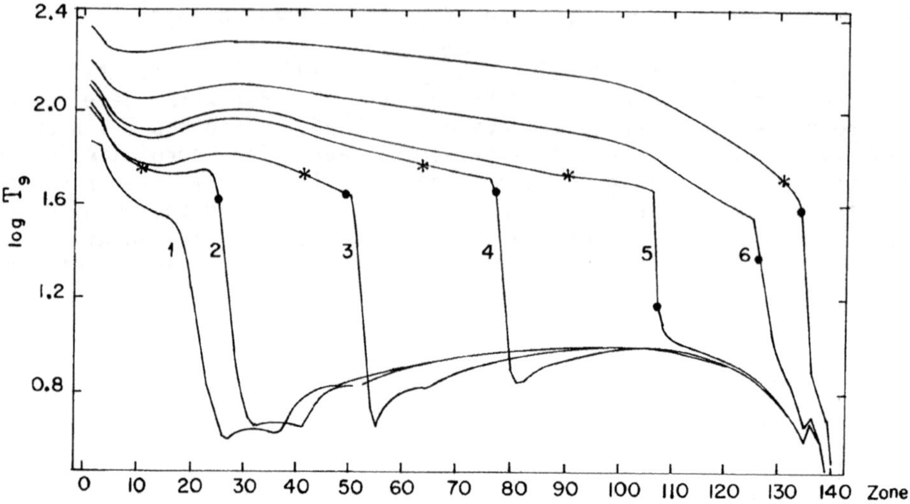

Figure 2. Temperatures versus Lagrangian coordinates in collapsing iron-oxygen star of 2 $M_\odot$ for the same moments of time as in Figure 1. The designations are the same as in Figure 1.

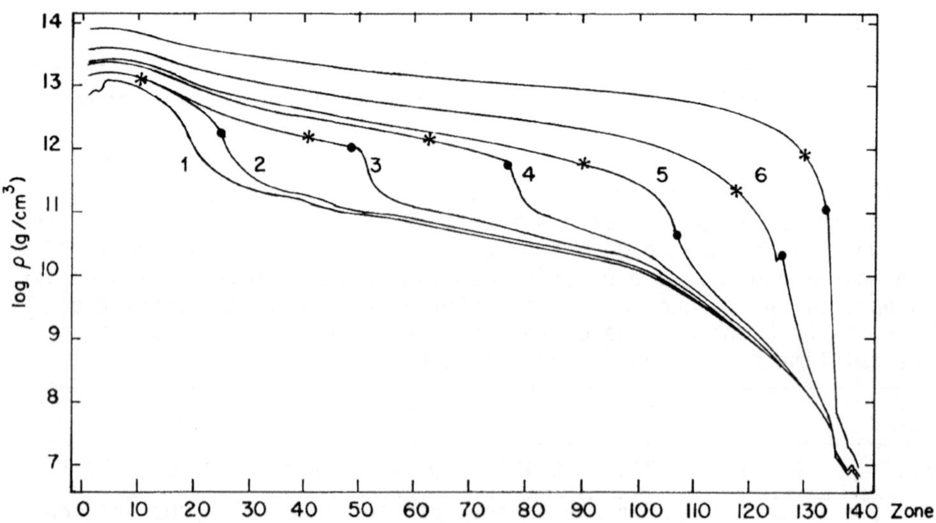

Figure 3. Densities versus Lagrangian coordinates in collapsing iron-oxygen star of 2 $M_\odot$. The designations are the same as in Figure 1.

The star 2 $M_\odot$ has collapsed in a hot neutron star. The joint action of the neutrino deposition and oxygen burning in the external layers, with the total mass 0.18 $M_\odot$, did not lead to the envelope ejection.

Energy deposition in the outer layers of the collapsing star from the neutrino photosphere, surrounding the nontransparent with respect to neutrino radiation stellar core, is of the same order of magnitude as the volume energy losses at the expense of their own neutrino radiation.

As for oxygen, it burns gradually: one layer burns after the other as it falls on the nontransparent core. The main part of energy is radiated at the stage of existence of the nontransparent collapsing core. Approximately the same number of neutrino of each sort (ν_e, $\tilde{\nu}_e$, ν_μ and $\tilde{\nu}_\mu$) are radiated since, in accordance with neutral currents

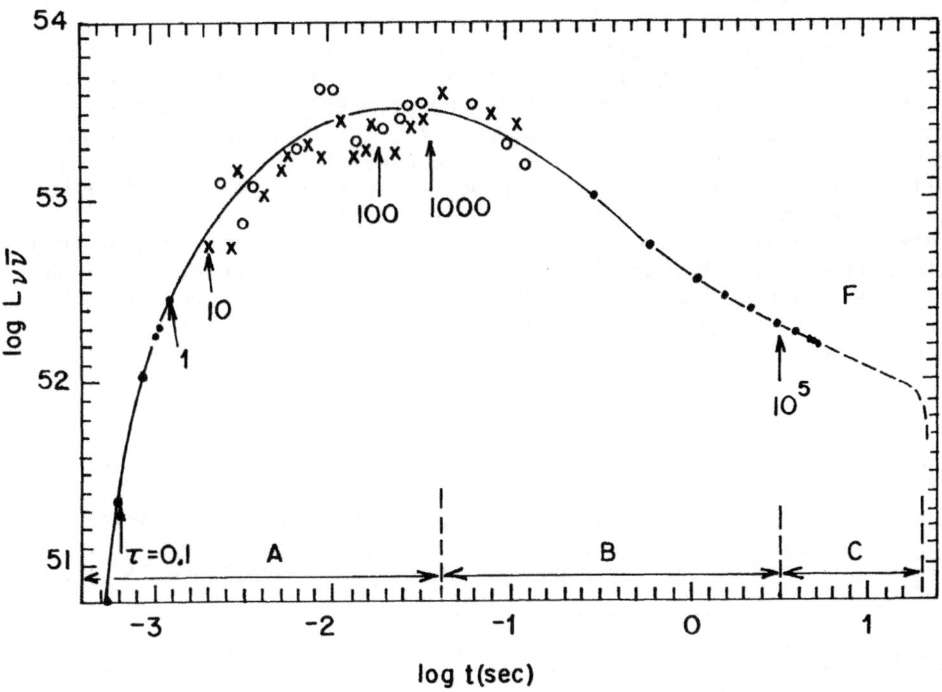

Figure 4. Total neutrino light curve for iron-oxygen star of 2 $M_\odot$ The time is measured from the moment when $t_{\nu\tilde{\nu}}$ = 0.001. The values of $t_{\nu\tilde{\nu}}$ are shown at the vertical arrows. The calculations were finished at moment denoted by F. The circles and the crosses correspond to various assumptions of muonic-neutrino radiation. Their dispersion is due to both the pulsations of the core and the computational errors. Capitals A, B and C denote the various phases of the collapse (see text, section 2).

of weak interactions, there is no significant difference between cross sections of electron- and muon-neutrino interactions with stellar matter (Weinberg, 1974).

A total neutrino light curve is presented in Figure 4. The collapse may be conditionally divided at the following three phases: A - the formation of neutrino opaque stellar core; B - the accretion of the envelope, delayed in collapse, onto the core; C - the Kelvin's cooling of a hot hydrostatically equilibrium neutron star due to neutrino energy losses. The characteristics of neutrino emission for the three phases are listed in Table 1. Figures 5-7 show the various parameter of neutrino opaque stellar core and of neutrino photosphere versus time, during the phase A and for the beginning of the phase B. The neutrino radiated spectrum is like the Fermian one with zero neutrino chemical potential (see Figure 7) and with photospheric temperature T_{ph} (Figure 6). The total neutrino luminosity, including all sorts of neutrinos, has a maximum about $3 \cdot 10^{53}$ ergs s^{-1} within characteristic time of $\sim$ 0.3 s (Figure 4). Taking into account the prolonged ($\sim$ 15 s) Kelvin cooling stage of the hot neutron star, the total energy about $5.8 \cdot 10^{53}$ ergs ($\sim$ 16 pc Mc) is radiated in the form of all sorts of neutrino. The mean neutrino particle energy is about 12 MeV in the vicinity of neutrino luminosity maximum. It is approximately keeping the same value throughout the phases A, B and C, when the gravitational reddening is taken into account (in Table 1 neutrino particle energy is given without allowance for gravitational reddening). Figure 8 shows the distributions of neutrino optical depth $t_{\nu\tilde{\nu}}(r)$ and of neutrino luminosity $L_{\nu\tilde{\nu}}(r)$ in Eulerian coordinates over the hydrostatically equilibrium hot neutron star, which was obtained at the end of calculations (point F in Figure 4). The neutrino luminosity derives the most of its energy from the gravitational energy release due to contraction of the shell with 7.5 $< r <$ 11 km.

Table 1

The characteristics of neutrino radiation for iron-oxygen 2 $M_\odot$ star.

$\mathcal{E}_{\nu\tilde{\nu}}$ - total energy radiated in the form of neutrino of all sorts.
$\overline{E}_{\nu\tilde{\nu}}$ - mean particle neutrino and antineutrino energy in MeV.
$\Delta t_{\nu\tilde{\nu}}$- duration of the corresponding phases of the collapse in sec.

Collapse Phase (Fig. 4)	A	B	C
$\mathcal{E}_{\nu\tilde{\nu}}$	1	17	30
$\overline{E}_{\nu\tilde{\nu}}$	12	14	15
$\Delta t_{\nu\tilde{\nu}}$	0.04	3.1	15

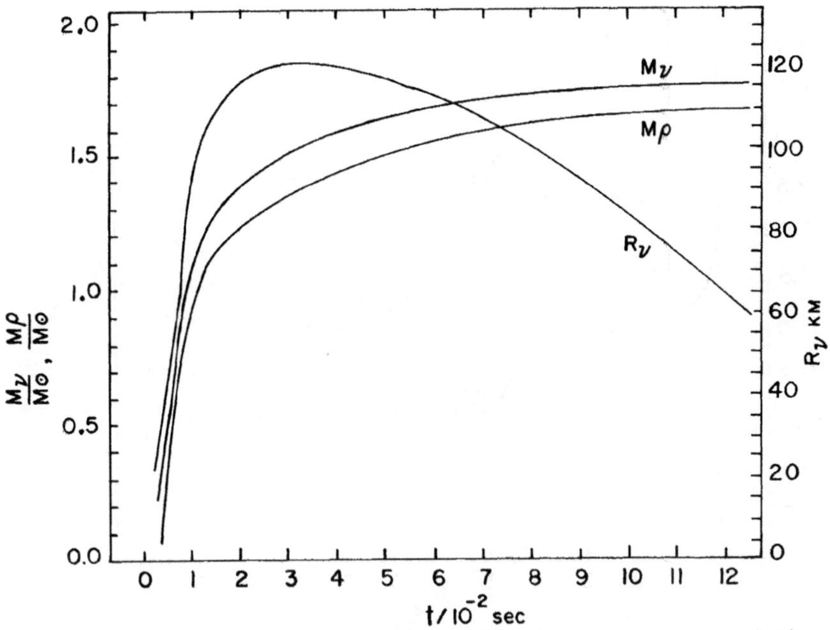

Figure 5. The mass M_ν and the radius R_ν of opaque core and the mass M_{ph} at the level of neutrino photosphere versus time for collapsing iron-oxygen star of 2 $M_\odot$.

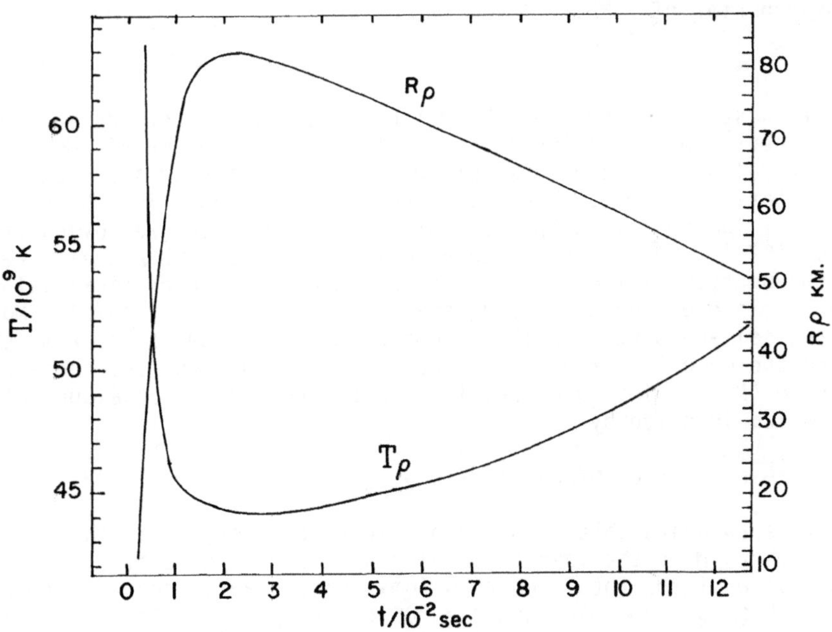

Figure 6. The radius R_{ph} and temperature T_{ph} at the level of neutrino photosphere versus time for collapsing iron-oxygen star of 2 $M_\odot$.

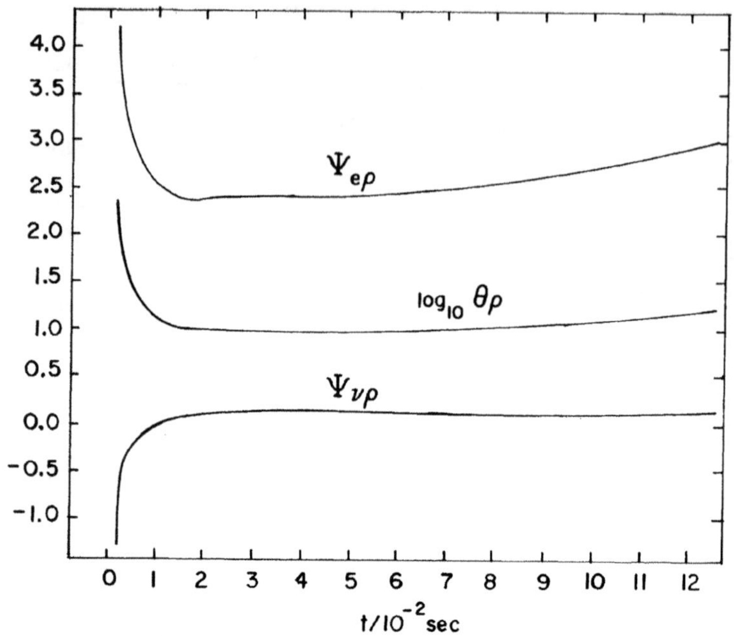

Figure 7. The chemical potentials of electrons Ψ_e and of neutrinos Ψ_ν (in units of k T) and the ratio of neutron and proton densities $\theta = n_n/n_p$ at the level of neutrino photosphere versus time for collapsing iron-oxygen star of 2 $M_\odot$.

With respect to neutrino absorption in the envelope, the maximum value of neutrino luminosity is at least two orders of magnitude less than critical one, in the sense of neutrino radiation pressure exceeding the gravitational force. There is here a full analogy with the notion of critical luminosity in the case of photon radiation (Zeldovich and Novikov, 1971). For coherent neutrino scattering on the nuclei like iron, the neutrino luminosity also does not exceed the critical value in the envelope of the collapsing star (Gershtein et al., 1975). Moreover, the self-consistent calculations of the collapse of the 2 $M_\odot$ star, counting the momentum transfer by means of coherent neutrino-nuclear scattering (Nadyozhin, 1976) resulted in a condition of envelope outburst, which is given by

$$\bar{A}\, a_0^2 > 1500\ (M/M_\odot)\ (10^{53}/L_{\nu\tilde{\nu}max}) \quad (1)$$

where a_0 is an adjustable constant of neutrino theory, $\bar{A}$ is the mean atomic number, M is the mass of a collapsing star, and $L_{\nu\tilde{\nu}max}$ is the maximal value of the total neutrino luminosity. Reference to (1) with $M = 2\ M_\odot$, $L_{\nu\tilde{\nu}max} = 3 \cdot 10^{53}$, and $\bar{A} = 100$ gives $a_0^2 > 10$. This is $\sim$ 50 times the value, adopted in Weinberg's theory, and is more than 10 times the lower limit, obtained by Wilson (1974).

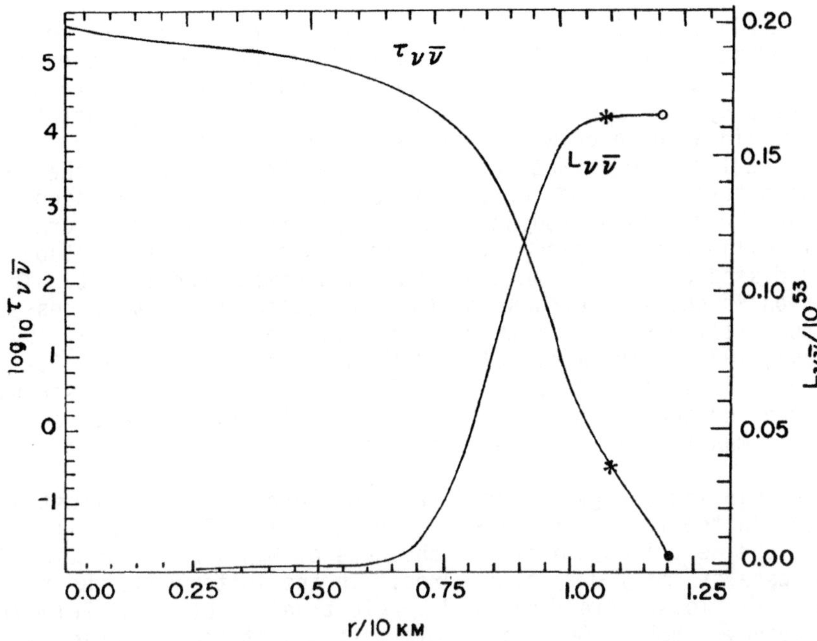

Figure 8. The distributions of the mean neutrino-antineutrino optical depth $t_{\nu\tilde{\nu}}$ and the total neutrino luminosity $L_{\nu\tilde{\nu}}$ along the hydrostatically equilibrium hot neutron star at the last moment of calculations (point F in Fig. 4). Marker · stands for the boundary of the opaque core and marker * for the position of neutrino photosphere.

The collapse of an iron-oxygen star 10 $M_\odot$ turns quickly into general-relativistic one and results in formation of a black hole without any hope of the envelope ejection (Nadyozhin, 1975b). The total energy release in this case is of $\sim 1.4 \cdot 10^{54}$ ergs (8 pc Mc^2) with the approximately same mean neutrino particle energy (12 MeV).

Barkat et al. (1975) have investigated the burning of various types of nuclear fuels against the background of the collapse and have also drawn a conclusion about the lack of ejected envelope.

Thus, the recent calculations all indicate the lack of the envelope ejection when iron cores collapse, i.e. neither the energy and momentum transfer from the collapsing core to envelope at hydrodynamic timescale ($\sim$ 1 s), nor the nuclear fuel burning in envelope is adequate for explanation of supernovae.

3. THE ONSET OF INSTABILITY IN IRON STARS WITH 1.2 $M_\odot$ < M_{Fe} < 2 $M_\odot$

The nonequilibrium neutronization is of a great importance for this mass interval. Basko and Imshennik (1975) have studied the development of hydrodynamic instability for iron stellar cores with masses M_{Fe} = 1.21, 1.4, and 1.7 $M_\odot$. The approach to the loss of stability has been considered by Sugimoto (1970) under the assumption of hydrostatic equilibrium. In accordance with Basko and Imshennik (1975), the iron stars of 1.4 and 1.7 $M_\odot$ contract, at first, very slowly owing to the nonequilibrium beta-process and the energy losses by neutrino emission. However, an abrupt (in hydrodynamic timescale) transition into gravitational collapse takes place at the moment when the loss of stability due to dissociation of iron-group nuclei becomes predominant. The dissociation is possible because of the nonequilibrium beta-processes are able to heat the stellar matter (Bisnovatyi-Kogan and Seidov, 1970; Nakazava, 1973). The model of 1.21 $M_\odot$ behaves differently. Its transition to collapse is gradual. The collapse develops in full strength only when the decreasing timescale of nonequilibrium beta-processes becomes comparable with the hydrodynamic one.

The binding energy of the resulting neutron star is almost all radiated in the form of neutrinos. There is a qualitative difference between the neutrinos emitted in the case of $M_{Fe} > 2\ M_\odot$ (section 2) and in the case of small mass iron cores. In the latter case the total neutrino energy losses are dominated by electronic neutrinos (Zeldovich and Guseinov, 1965). Only at the last stages of the collapse, just before neutron star formation, some energy may be radiated with the help of electronic antineutrinos. The energy losses through muonic neutrinos and antineutrinos are negligible for all stages of the collapse.

There is not any quantitative investigation of envelope outburst for the masses of iron cores discussed in this section.

4. CARBON FLASH DYNAMICS

The investigation of carbon flash is very complicated because of the necessity to take into account the kinetics of (a) nuclear burning in the degenerated stellar matter, and of (b) neutronization of carbon burning products. The pioneer investigations in this field were fulfilled by Barkat et al. (1970) and Bruenn (1972). Ivanova et al. (1974, 1975) carried out the self-consistent computations of both these effects and got the following results:

1) The initial temperature distribution is of importance to hydrodynamic development of the carbon flash. It was taken in the form to reproduce as close as possible the final result of the prolonged (10^5-10^6 sec) thermal phase of the carbon flash, which is controlled by convection and heats the matter up to $T \approx 8 \cdot 10^8$ K in a small central region of the carbon core. The dependence of basic results on the initial temperature distribution has been investigated. Three initial temperature profiles, used in calculations for model with initial central density of $2.3 \cdot 10^9$ g cm^{-3}, are plotted in Figure 9.

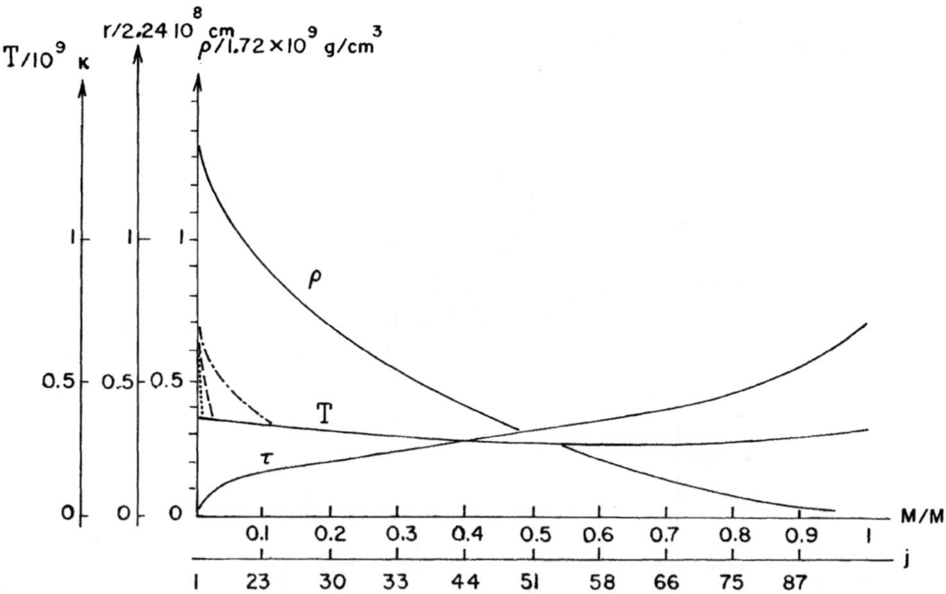

Figure 9. Initial structure of the hydrostatistically equilibrium carbon core: densities ρ, temperatures T, and Eulerian radii r versus Lagrangian mass m/M and mesh point number j. The mass of carbon core M = 1.40 $M_\odot$, the initial central density $\rho_c = 2.34 \cdot 10^9$ g cm^{-3}, and the initial central temperature $T_c = 3.69 \cdot 10^9$ K. The dashed, dotted, and dash-dotted curves in the vicinity of the center indicate temperature distributions, with the same value $T_c = 8 \cdot 10^8$ K at the center, which imitate preceding thermal stage of carbon flash.

2) The carbon flash evolves in a deflagrative regime, although the calculations do not take into account any mechanisms of propagation of burning (convection, thermal conductivity) except of hydrodynamic compression. The quasi-stationary detonative burning is of a small probability.

3) The deflagration results in the increasing radial pulsations of the carbon core. After combustion of some carbon fuel, there is either the total disruption of a star or the collapse of a stellar core, depending on the value of initial central density and on the initial temperature distribution. When a very narrow central region is initially heated up, the almost hydrostatic development of the carbon flash is not excluded. In this case the iron core is ultimately formed and then evolves in accordance with the scenario discussed in section 3. The calculations have not resulted in simultaneous formation of the collapsed core and the ejected envelope (see Figures 10-15).

Nomoto et al. (1976) took account of convection and also found out the deflagrative character of the carbon burning.

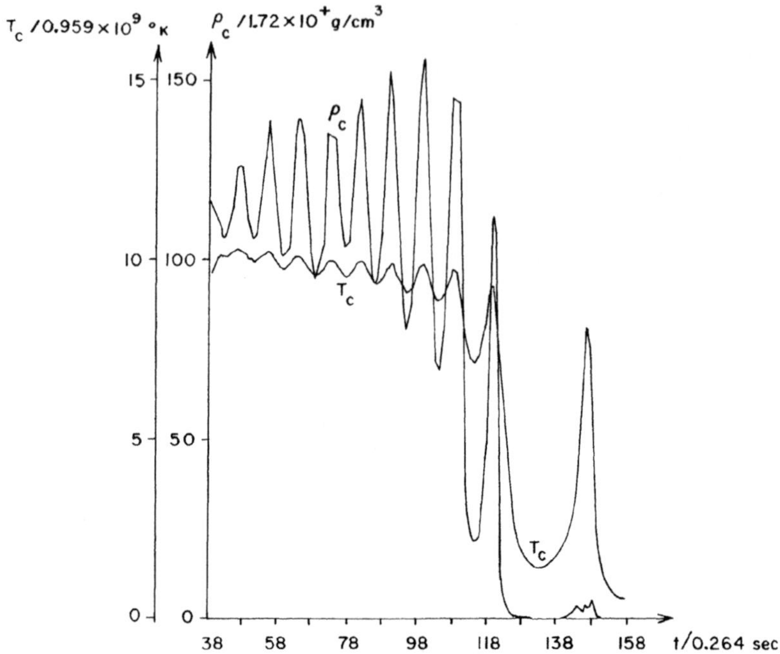

Figure 10. Pulsation regime of the thermonuclear carbon burning. Central temperature T_C and central density ρ_C versus time for dashed initial temperature distribution (Fig. 9). The dimensions of region, where temperature was increased above its hydrostatic value, are $1 \leq j \leq 15$, $0 \leq m/M \leq 0.029$. After ten pulsations the core is totally disrupted.

4) For initial central density $\rho_C < (3\text{-}5) \cdot 10^9$ g cm^{-3} the star is completely disrupted without giving any remnant of the neutron star type. However, the energy of expansion is insufficient to explain the supernova outbursts. The fact is that the carbon fuel is consumed only in the vicinity of stellar center (see Fig. 13) in contrast to Arnett's (1969) paper where, due to artificial stimulating of detonative burning regime, the carbon was completely consumed.

5) In the case of $\rho_C > (3\text{-}5) \cdot 10^9$ g cm^{-3} the carbon flash leads to the gravitational collapse of the stellar interiors owing to neutronization of carbon burning products and to intensive neutrino energy losses.

The decrease of entropy during the collapse of carbon burning products is considerably greater in comparison with the collapse of iron-oxygen stellar cores. It manifests itself in a low minimum of central temperature ($\sim 10^9$ K) at the moment when the density of the most intensive neutronization ($\sim 10^{12}$ g cm^{-3}) is attained (see Fig. 16).

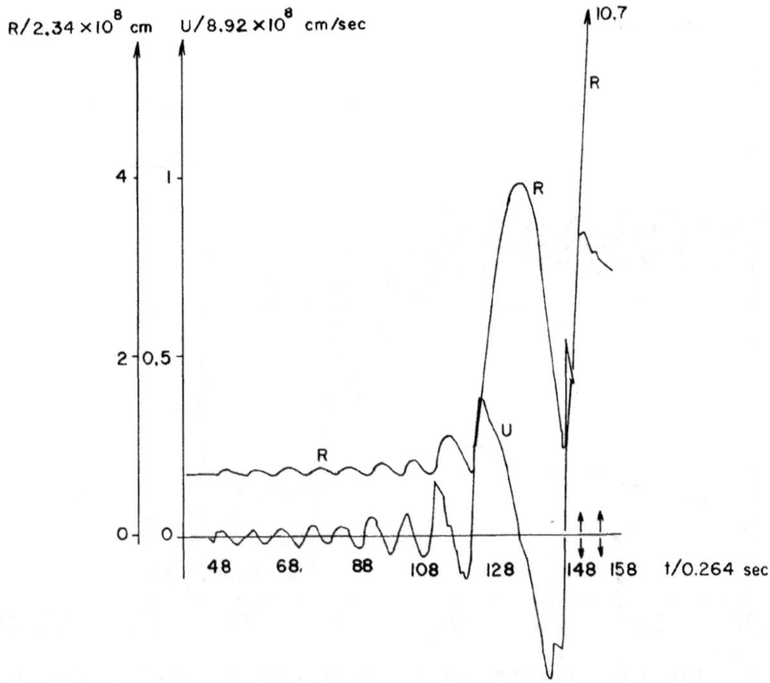

Figure 11. Pulsation regime of the thermonuclear carbon burning. The velocity u and the radius R of external surface of the core versus time. The initial conditions are the same as in Fig. 10. Moments of time both when velocity becomes greater than escape one only for the external mesh point and all over the star are marked by arrows on time-axis.

Thus, the critical value of central density is $(\rho_c)_{crit} \underset{\sim}{\sim} (3\text{-}5) \cdot 10^9$ g cm^{-3}. According to the theory of evolution of the single stars, central density just before onset of the carbon flash is $\rho_c = (2\text{-}3) \cdot 10^9$ g cm^{-3} independently on the mass of a star at the main sequence. In the theory of evolution of close binaries, however, the higher central density values $\rho_c \gtrsim 5 \cdot 10^9$ g cm^{-3} are quite possible (Ergma and Tutukov, 1976).

For $\rho_c > 5 \cdot 10^9$ g cm^{-3}, alongside with the collapsing stellar interiors, there is a possibility of envelope ejection with the energy adequate for supernova outburst. The main triggering mechanism of envelope ejection in this case is the neutrino deposition which initiates the ignition of carbon in detonative regime (Gershtein et al., 1975). A considerably higher efficiency of the deposition for ignition of nuclear fuel, as compared with the collapse of iron stars, is due to the following reasons:

1) Owing to a relativistic degeneracy of electrons, the cross

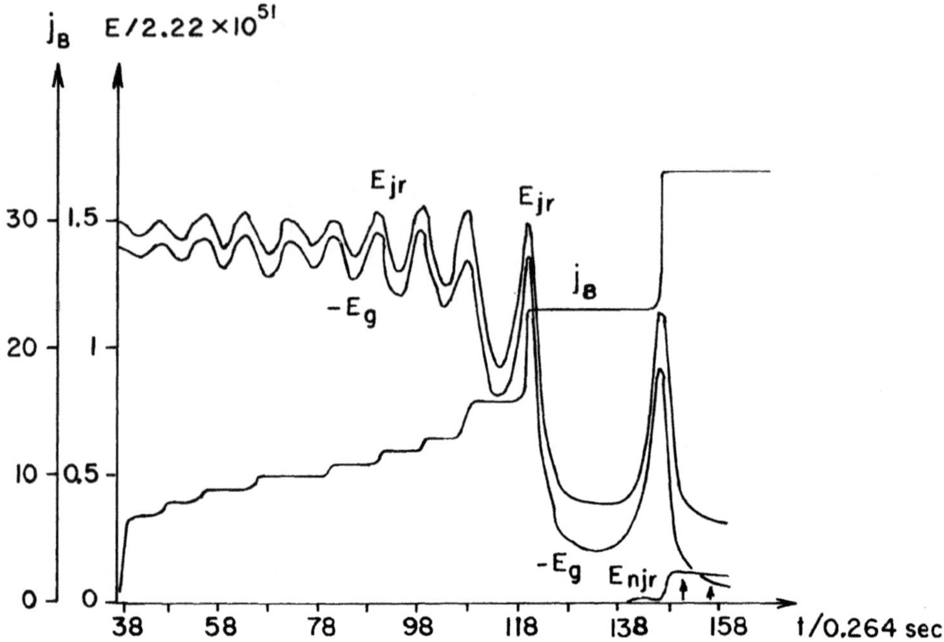

Figure 12. Integral energies of all star versus time for the same initial conditions as in Figures 10 and 11. Internal energy E_{in} includes the rest energy of electrons E_e ($E_{eo} = 6.86 \cdot 10^{50}$ ergs for $t = 0$). E_g is gravitational energy, and E_{kin} is kinetic energy. Initial total energy of the star equals $E_o = -5.1 \cdot 10^{50}$ ergs when E_{eo} being subtracted. The total energy at the end of calculations equals $E_f = 1.3 \cdot 10^{50}$ ergs. The number of burned mesh points j_b is also plotted out.

section of neutrino-electron scattering (in this case the scattering is the main mechanism of neutrino interaction with stellar matter) increases by one order in magnitude since cross section is proportional to electron chemical potential.

2) The heat capacity of degenerate matter is very small and, therefore, the energy, necessary for heating of the matter up to carbon ignition temperature ($T_{ign} \approx 10^9$ K), is also small.

3) The ratio of the radius of effective region, where the neutrino radiation is produced at the expense of nonequilibrium beta-processes with products of carbon burning, to the envelope radius is considerably greater than for collapsing iron-oxygen stars. As a result the neutrino radiation is not so strongly diluted and the volume neutrino density is higher in the envelope.

The self-consistent calculations of the stellar core collapse,

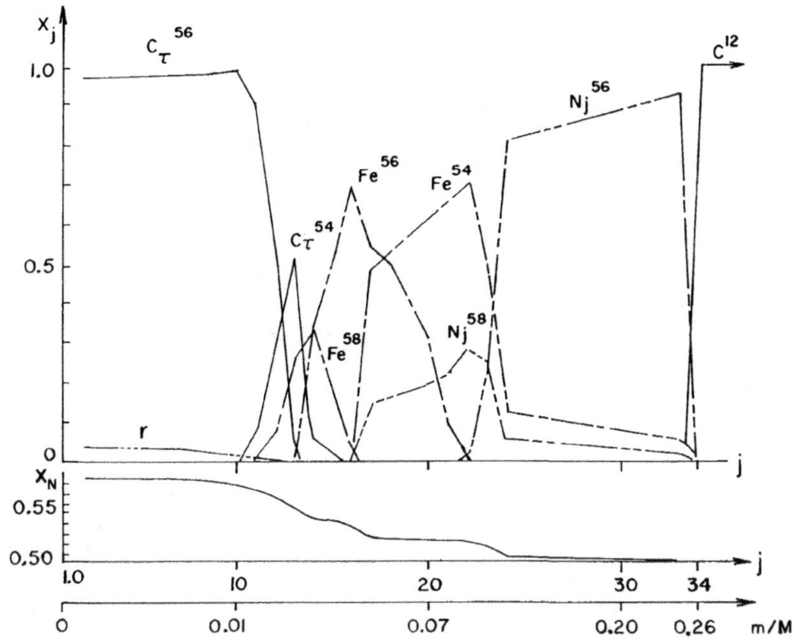

Figure 13. Abundances X_j by weight of various nuclei and free neutrons versus j for one of the last moments of time t = 149.7 (or t = 39.52 s in physical units). The total abundance of neutrons X_N (free plus bound) is also given at the bottom of the figure. Neutronization is maximal at the center where X_N = 0.565. Carbon has all burned from the center to j = 34, total burned mass being $\Delta m/M \approx 0.25$. The initial conditions are the same as in Figures 10-12.

alongside with neutrino-induced carbon burning in the envelope, may additionally result in an increase of the critical density $(\rho_c)_{crit}$ value since the ignition accelerates the carbon burning. These calculations are now underway.

4. SUPERNOVA OUTBURSTS AND SOME PROBLEMS OF NUCLEOSYNTHESIS

The theory of advanced stages of stellar evolution is believed to account for two basic observational facts: (1) supernova outbursts, and (2) the abundances of chemical elements in Galaxy. Supernova outbursts are usually supposed to be connected with emergence of a powerful shock wave into a stellar envelope. Grassberg and Nadyozhin (1969) and Grassberg et al. (1971) have shown that supernova light curves can be reproduced in that case if only a presupernova is surrounded by extended envelope of dimensions which are comparable with the observed photospheric radii (5,000-100,000 $R_\odot$) at supernovae maximum light.

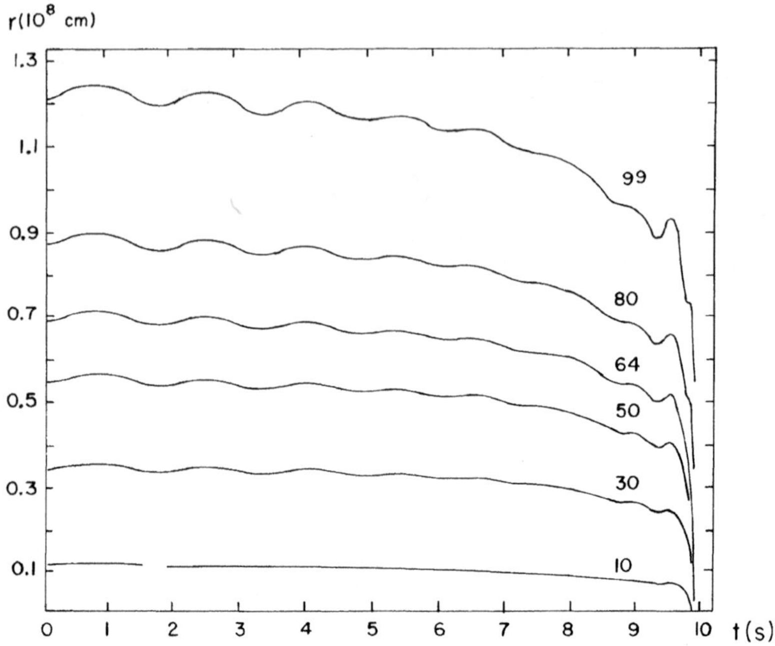

Figure 14. Eulerian radii of various mesh points versus time. Initial central density equals $\rho_c = 5.05 \cdot 10^9$ g cm^{-3} and total stellar mass is $M = 1.40\ M_\odot$. The curves are labeled by mesh point number j which corresponds to the same values of $\Delta m/M$ as in Figures 9 and 13. After some pulsations the star all begins to collapse at t $\gtrsim$ 9.9 s.

The amazing stability of the collapsing iron-oxygen stars with respect to envelope ejection brings in evidence a small probability of the powerful shock wave generation in that case. Therefore, though there is a kind of hope to generate a shock wave in the process of carbon flash, some other mechanisms of envelope ejection deserve consideration. In the first place, rotation and magnetic field should be considered. Some effects, connected with rotation and magnetic field were discussed by Ivanova et al. (1969) from theoretical point of view and by Mustel (1970) and Shklovski (1970) from observational one. Le Blanc and Wilson (1970) have performed very complex calculations counting the interaction of rotation with superhigh magnetic field against background of the gravitational collapse. This numerical experiment points to a possibility of envelope ejection in that case.

Bisnovatyi-Kogan (1970) and Bisnovatyi-Kogan et al. (1976) investigated the envelope ejection due to transfer of angular momentum from collapsing core to envelope by means of magnetic field. This process can proceed for a timescale exceeding the hydrodynamic one ($\sim$ 1 s). The interaction of magnetic-dipole radiation from newborn pulsar with un-

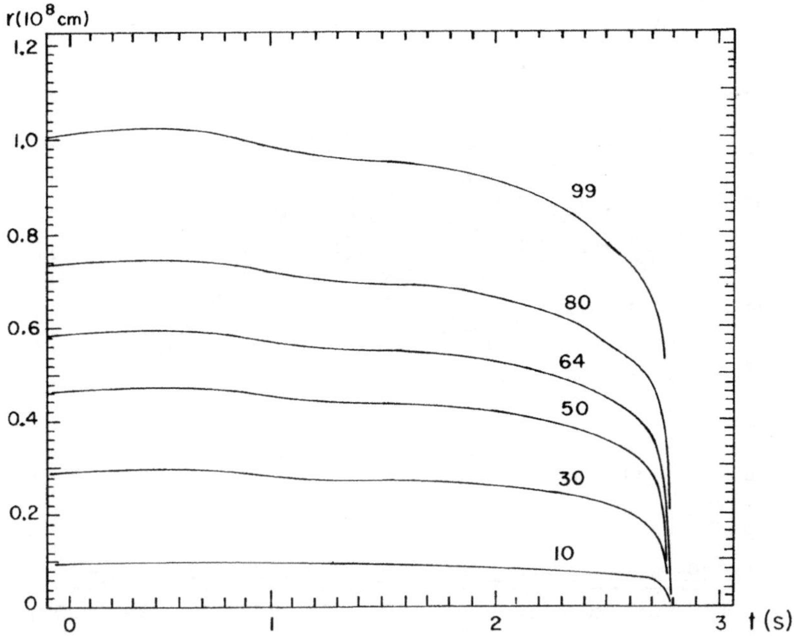

Figure 15. Eulerian radii of various mesh points versus time for variant which differs from variant, plotted in Fig. 14, only by initial central density: $\rho_c = 8.36 \cdot 10^9$ g cm^{-3}. The collapse begins at t $\gtrsim$ 2.8 s. The pulsations are hardly noticeable.

collapsed envelope was investigated by Ostriker and Gunn (1971) and Bodenheimer and Ostriker (1974).

Thus, at the present time there is a good background of evidences for further development of the new point of view on the supernova outbursts, which deals with rather slow expelling of envelope but not with powerful shock wave.

A modelling of supernova outbursts with the help of slow pumping of energy (within a timescale exceeding the hydrodynamic one but not so high as thermal one) into an envelope was undertaken by Nadyozhin and Utrobin (1975, 1976). Using the set of hydrodynamic equations, with radiative heat conductivity taken into account, they found out that the observational data for supernovae of type I are well reproduced when the envelopes of masses (0.3-0.6) $M_\odot$ are supplied with total amount of energy (2-5) 10^{50} ergs during a time of (2-20) days (Figure 17).

The applicability of radiative heat conductivity to the supernova problem requires special examination. As the matter in supernova envelope is strongly ionized, the probabilities of photon's absorption

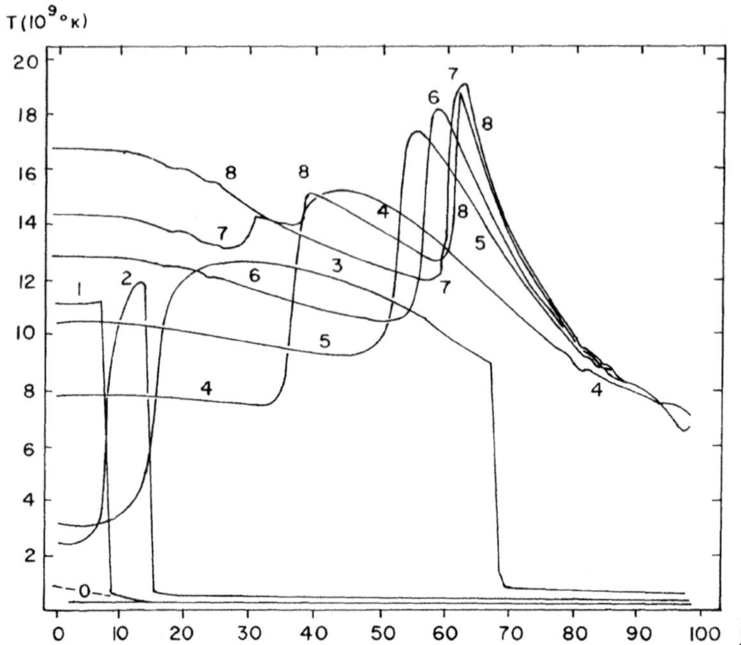

Figure 16. Temperatures versus j for various moments of time which are labeled by integers. The correspondence between time in seconds and integers is given by: 0-0, 1-0.3165, 2-2.433, 3-2.725, 4-2.7714, 5-2.77817, 6-2.77917, 7-2.77997, 8-2.78017. Initial central density $\rho_c = 8.36 \cdot 10^9$ g cm^{-3} (see Fig. 15) and stellar mass $M = 1.4\ M_\odot$. A front of carbon burning is well noticeable up to the moment 4 when carbon all burned out over the star. One can see the central region which is cooled by neutrino energy losses. The minimum temperature value of $\sim 2 \cdot 10^9$ K takes place at moment 2. Total increase of temperature for moments 4-8 is owing to compression of collapsing matter. The broken bends of curves for moments 7 and 8 are due to a crude counting of neutron degeneracy.

and emission are to a considerable extent smaller than the probability of electron scattering. In that case the radiative heat conductivity approximation seems to be very crude one because of the local thermodynamic equilibrium between matter and radiation being violated. For correct treating of photon's diffusion in the supernova envelope, the approximation of so-called Compton radiative gas dynamics (CRG) should be used (Imshennik, 1976). With the density increasing, this approximation turns asymptotically into the radiative heat conductivity (RHC). In the CRG-theory photons are distributed in accordance with Bose-Einsteinian law. The number density of photons is now a new independent function of space and time. For an extended model ($R_0 = 10^4\ R_\odot$) of supernovae of type II ($M_0 = 17\ M_\odot$) a light curve was calculated as a result of artificial explosive release of energy in amount of E_0 =

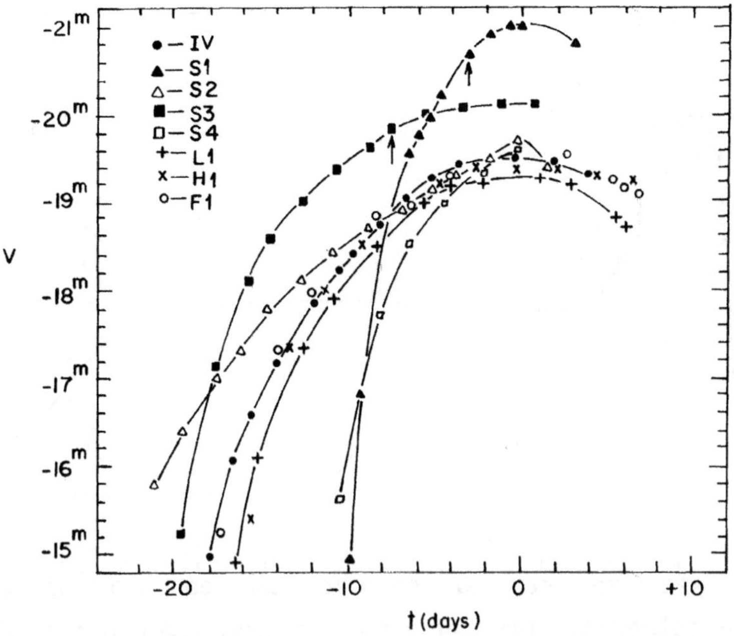

Figure 17. Theoretical V-light curves (in UBV-system) for supernova models of type I with slow energy pumping. Maxima of the light curves are reduced to the same time. The mass of ejected envelope is the same for all variants and equals to 0.45 $M_{\odot}$. Variants IV, S2, S4, L1, H1, and F1 are characterized by approximately the same energy release of $\sim 3 \cdot 10^{50}$ ergs but the time of energy pumping being varied from 2 to 20 days. The stellar models all consist of pure helium, except variant H1 for which helium was replaced by pure hydrogen. Energy release for variant S1 is of $2.2 \cdot 10^{51}$ ergs and for S3 is of $5.6 \cdot 10^{50}$ ergs. Vertical arrows for S1 and S3-curves point to the position of bolometric maxima.

$3 \cdot 10^{50}$ ergs (Imshennik and Utrobin, 1976). The calculations were fulfilled both in the CRG and in the RHC theories. The difference between the two theories is very small, as bolometric light curves concerns (Figure 18). However, there is an essential divergence of temperature distributions inside supernova envelope at the moment of maximum light (Figure 19). The deficiency of photon number density is also quite obvious in Figure 19. In compact stellar models relatively high temperatures are attained, and the divergences from Planckian distribution law may be, therefore, more significant.

When the stellar envelope is slowly expelling, the matter may be for a long time in the neighborhood of collapsed central remnant. Under such conditions the interaction of neutrinos, radiated by collapsing stellar core and then by hot neutron star (Figure 4), with the envelope

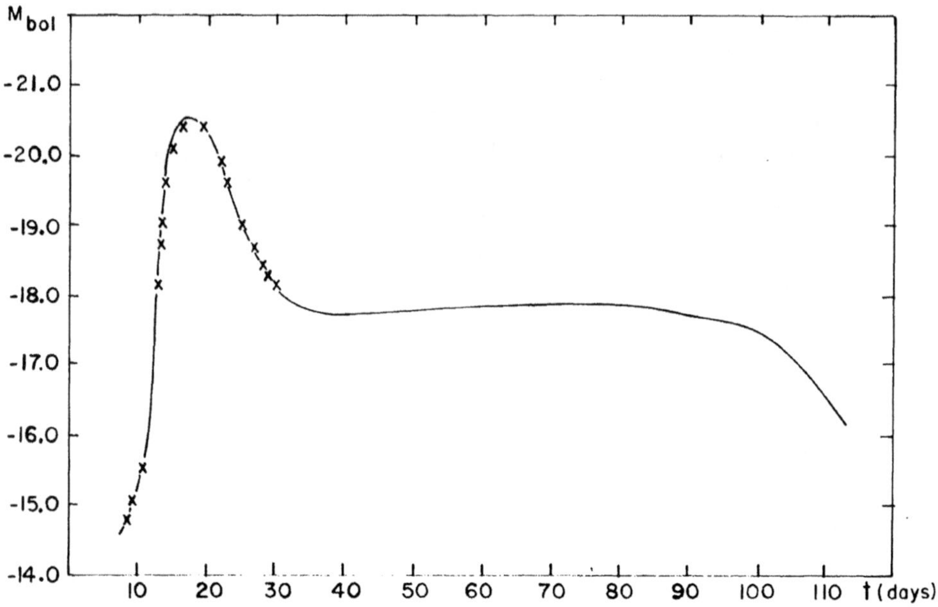

Figure 18. A bolometric light curve for supernova model of type II with extended hydrogen envelope. The full curve has been calculated with the help of radiative heat conductivity (RHC) approximation and crosses stand for calculations within Compton radiative gas dynamics (CRG), see text (section 4) for details.

may be of importance for some problems of nucleosynthesis. Domogatsky and Nadyozhin (1976) exposed the matter, consisting of stable isotopes in natural mixture, to the neutrino flux from the collapsing stellar core (section 2, Figure 4). They found out that the most of proton-rich (bypassed) nuclei are well produced in accordance with their observed cosmic abundances when the matter is disposed at the distances of (0.3-1) 10^9 cm from the center for a time which is not too less than the duration of neutrino light curve ($\sim$ 15 s, see Figure 4 and Table 1).

When rotation and magnetic field are neglected, from the above discussion it follows that the simultaneous origin of collapsed remnant (neutron star) and supernova outburst is only possible on the way of the carbon flash investigation. The ejected envelope in the case has to be enriched mainly with carbon burning products: alpha-nuclei, iron-group elements, and possibly some r-process elements. This model does not fit to the existence of nuclei heavier than iron-group ones, to say nothing about superheavy elements ($Z \gtrsim 126$), discovered just now (Gentry et al., 1976). The heavy nuclei can be synthesized in undersurface layers of neutron stars (Chechetkin, 1971; Bisnovatyi-Kogan and Chechetkin, 1974). The enrichment of interstellar

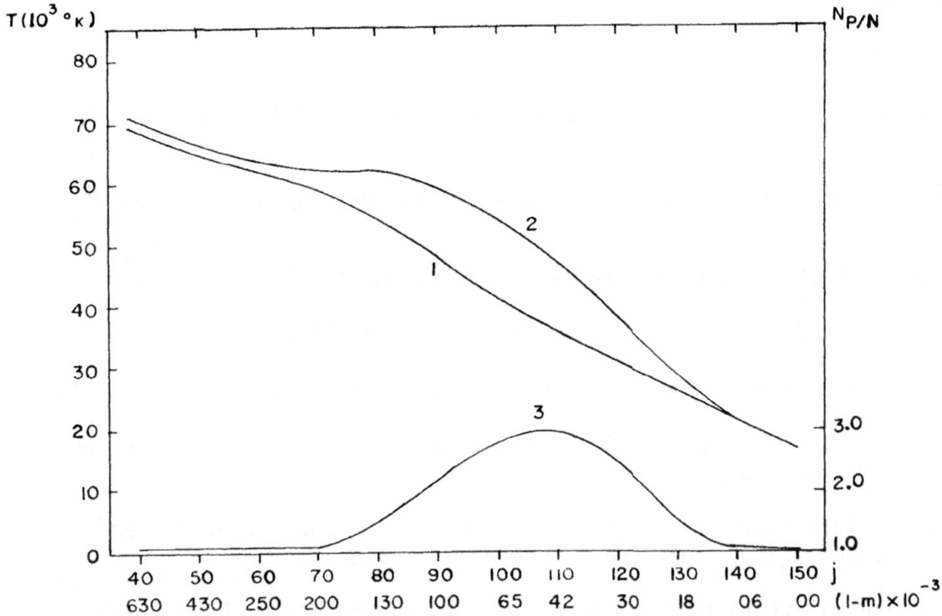

Figure 19. Temperature versus mesh point number j or Lagrangian relative mass q measured from stellar surface, q = (M - m)/M, for the moment of bolometric maximum (17.5 days, Figure 18). The curve 1 is for RHC and the curve 2 is for CRG-calculations. The curve 3 gives a ratio of Planckian photon density N_p, calculated for temperature curve 2, to real photon density N derived from CRG-calculations. An excess of temperature and deficiency of photon density are well noticeable for CRG-calculations as compared with RHC ones.

medium with heavy nuclei in this case is performed by means of numerous ejecta of the matter from the neutron stars which may be followed by observed jumps in the periods of the pulsars. Bisnovatyi-Kogan et al. (1975) found that the ejecta may manifest itself in the form of observed pulsed gamma-ray emission of cosmic origin. The evolution of the matter, ejected from neutron stars, was qualitatively considered by Chechetkin and Kowalski (1976). A mechanism of matter ejection from chemically nonequilibrium undersurface layers of neutron star was discussed by Bisnovatyi-Kogan et al. (1976).

REFERENCES

Arnett, W. D. 1969, Astrophys. Space Sci. 5, 180.

Barkat, Z., Buchler, J. R., and Wheeler, J. C. 1970, Astrophys. Letters 6, 117.

Barkat, Z., Rakavy, G., Reiss, Y., and Wilson, J. R. 1975, Astrophys. J. 196, 633.

Basko, M. M., and Imshennik, V. S. 1975, Astron. Zh. 52, 469.

Bisnovatyi-Kogan, G. S. 1970, Astron. Zh. 47, 813.

Bisnovatyi-Kogan, G. S., and Seidov, Z. F. 1970, Astron. Zh. 47, 139.

Bisnovatyi-Kogan, G. S., and Chechetkin, V. M. 1974, Astrophys. Space Sci. 26, 25.

Bisnovatyi-Kogan, G. S., Imshennik, V. S., Nadyozhin, D. K., and Chechetkin, V. M. 1975, Astrophys. Space Sci. 35, 3.

Bisnovatyi-Kogan, G. S., Popov, Yu. P., and Samochin, A. A. 1976, Astrophys. Space Sci. 41, 287.

Bisnovatyi-Kogan, G. S., Kulikov, Y. N., and Chechetkin, V. M. 1976, Inst. Appl. Math. Acad. Sci. USSR, Moscow, Preprint No. 8.

Bodenheimer, P., and Ostriker, J. P. 1974, Astrophys. J. 191, 465.

Bruenn, S. W. 1972, Astrophys. J. Suppl. 24, 283.

Chechetkin, V. M. 1971, Astron. Zh. 48, 59.

Chechetkin, V. M., and Kowlaski, M. 1976, Nature 259, No. 5545, 643.

Colgate, S. A., and White, R. 1966, Astrophys. J. 143, 626.

Domogatsky, G. V., and Nadyozhin, D. K. 1976, Inst. Appl. Math. Acad. Sci. USSR, Moscow, Preprint No. 94.

Ergma, E. V., and Tutukov, A. V. 1976, Acta Astron. 26, 69.

Fowler, W., and Hoyle, F. 1965, Neutrino Processes and Pair Formation in Massive Stars and Supernovae (Chicago: University of Chicago Press).

Gentry, R. V., Cahill, T. A., Fletcher, N. R., Kaufmann, H. C., Medsker, L. R., Nelson, J. W., and Flocchini, R. G. 1976, Phys. Rev. Letters 37, 11.

Gerstein, S. S., Imshennik, V. S., Nadyozhin, D. K., Folomeshkin, V. N., Khlopov, M. Yu., Chechetkin, V. M., and Eramzhyan, R. A. 1975, JETP (USSR) 69, 1473. See also a short communication in Physics Letters 62B, 100 (1976).

Grassberg, E. K., and Nadyozhin, D. K. 1969, Astron. Zh. 46, 745.

Grassberg, E. K., Imshennik, V. S., and Nadyozhin, D. K. 1971, Astrophys. Space Sci. 10, 3.

Hansen, C. J., and Wheeler, J. C. 1969, Astrophys. Space Sci. 3, 464.

Imshennik, V. S., and Nadyozhin, D. K. 1972, JETP (USSR) 63, 1544. (Translation in English: 1973, Soviet Physics JETP 36, 821).

Imshennik, V. S., and Nadyozhin, D. K. 1974, Late Stages of Stellar Evolution, ed. Tayler, IAU (Dordrecht: D. Reidel Publishing Co.), 130.

Imshennik, V. S. 1976, Inst. Appl. Math. Acad. Sci. USSR, Moscow, Preprint No. 86.

Imshennik, V. S., and Utrobin, V. P. 1976, Astron. Zh. Pisma, in press.

Ivanova, L. N., Imshennik, V. S., and Nadyozhin, D. K. 1969, Nauch. Inform. Astron. Soviet. Acad. Sci. USSR 13, 3.

Ivanova, L. N., Imshennik, V. S., and Chechetkin, V. M. 1974, Astrophys. Space Sci. 31, 497.

Ivanova, L. N., Imshennik, V. S., and Chechetkin, V. M. 1975, Inst. Appl. Math. Acad. Sci. USSR, Moscow, Preprint No. 31.

Le Blanc, J. M., and Wilson, J. R. 1970, Astrophys. J. 161, 541.

Mustel, E. R. 1970, Astrophys. Letters 6, 207.

Nadyozhin, D. K. 1975a, Inst. Appl. Math. Acad. Sci. USSR, Moscow, Preprint No. 98.

Nadyozhin, D. K., 1975b, Inst. Appl. Math. Acad. Sci. USSR, Moscow, Preprint No. 106.

Nadyozhin, D. K., 1976, Inst. Appl. Math. Acad. Sci. USSR, Moscow, Preprint No. 26.

Nadyozhin, D. K., and Utrobin, V. P. 1975, Inst. Appl. Math. Acad. Sci. USSR, Moscow, Preprint No. 122.

Nadyozhin, D. K., and Utrobin, V. P. 1976, Inst. Appl. Math. Acad. Sci. USSR, Moscow, Preprint No. 85.

Nakazawa, K. 1973, Progr. Theor. Phys. 49, 1932.

Nomoto, K., Sugimoto, D., and Neo, S. 1976, Astrophys. Space Sci. 39, L37.

Ostriker, J. P., and Gunn, J. E. 1971, Astrophys. J. 164, L95.

Paczynski, B. E. 1970, Acta Astron. 20, 47.

Rose, W. K. 1969, Astrophys. J., 153, 491.

Shklovsky, I. S. 1970, Astrophys. Letters 8, 101.

Sugimoto, D. 1970, Astrophys. J. 161, 1069.

Uus, U. 1970, Nauch. Inform. Astron. Soviet. Acad. Sci. USSR 17, 3.

Weinberg, S. 1974, Rev. Mod. Phys. 46, 255.

Wilson, J. R. 1971, Astrophys. J. 163, 209.

Wilson, J. R. 1974, Phys. Rev. Letters 32, 849.

Zeldovich, Ya. B., and Guseinov, O. X. 1965, Doklady Acad. Sci. USSR 162, 791.

Zeldovich, Ya. B., and Novikov, I. D. 1971, Stars and Relativity, vol. 1, ed. Arnett and Thorne (Chicago: University of Chicago Press).

MECHANISMS FOR SUPERNOVA EXPLOSIONS

Richard I. Epstein
Center for Astrophysics, Harvard College Observatory and
Smithsonian Astrophysical Observatory, and
NORDITA, Copenhagen, Denmark

Considerable progress has been made in the study of the late stages of stellar evolution, gravitational collapse, and neutrino physics. However, the mechanisms by which stars undergo supernova explosions is still not understood. This report discusses some of the recent developments in the study of one supernova mechanisms, the neutrino transport mechanism, and indicates what future developments are needed before this model can be adequately understood.

Recent work on the later stages of stellar evolution (see Paczynski, 1970; Arnett, 1972) has greatly clarified the presupernova stellar configuration for isolated stars. All stars which are massive enough to permit the ignition of carbon under conditions of low electron degeneracy, that is stars of masses greater that 5-7 $M_\odot$, evolve so that the stellar cores converge to very similiar structures. The central burning core of each of these stars remains at near the Chandrasekhar mass limit as it goes through sucessive burning stages. Finally all the available nuclear fuel in the center is consumed and the core evolves to essentially a hot, iron, white dwarf of about 1.3 to 2.2 $M_\odot$. Because the core is considerable more dense than the surrounding stellar envelope, it is in many respects a separate dynamical entity. When a core collapses and possibly explodes, it does so on a timescale which is short compared to the timescale with which the outer regions of the star can respond; the outer regions of the star can only hover above the core during this brief interval.

As the iron core evolves it begins to contract and then to collapse. This explosion is triggered by two processes: electron capture reactions and photodisscociation reactions (Fowler and Hoyle, 1964; Hansen and Wheeler, 1969; Barkat, 1975). The electron captures tend to decrease the density of electrons and lessen the degeneracy pressure, the main support of the stellar core, and the endothermic dissociation of the heavy iron peak elements decreases the thermal pressure of the ions. For masses near 1.5 $M_\odot$ both processes proceed nearly simultaneously giving rise to a collapsing, hot central core composed largely of free neutrons surrounded by layers of first helium and

David N. Schramm (ed.), Supernovae. 183-192.

neutrons and then neutron rich iron peak elements.

Can the phenomena of collapse and implosion of the dense stellar core and the processes which accompany this implosion - the release of neutrinos and the change in the composition of the matter - produce an explosion which yeilds sufficient energy in the right form to generate a supernova event? That is, can the implosion of a iron white dwarf core in the center of a massive star transfer about 10^{51} ergs to the outer regions of the star?

Much work has been done recently on studying the hydrodynamics of the collapse and rebound of stellar cores and the roles which the production and transport of neutrinos play in these events. This work has largely been within the context of the Neutrino Transport model which was originally proposed by Colgate and White (1966). This model assumes a spherically symmetric collapse in which the increase in the binding energy of the core is the only source of energy (Wilson, 1971, 1974; Wilson et al., 1975; Bruenn, 1975; Schramm and Arnett, 1975). In an actually supernova event other factors, such as rotation, magnetic fields and ignition of nuclear fuels, may be also important (Fowler and Hoyle, 1964; Barkat et al., 1975; Le Blanc and Wilson, 1970; Meier et al., 1976). Nevertheless until the simpler, perhaps over idealized, problem of the neutrino transport supernova is solved one can have little hope of understanding supernova models which involve additional complications.

The basic picture of the current neutrino transport supernova can be described with the aid of figures 1 and 2. In these figures each line represents the position of an element of matter in a 1.5 $M_{\odot}$ stellar core as a function of the time which has elapsed since the start of the collapse. These figures are based on some unpublished calculations by J. Wilson. These two calculations differ in the physical parameter, such as the neutrino opacities, which were used. The particular input physics which was used in these two models is not relevant to the present discussion.

First consider figure 1. This is an example of a core collapse which cannot lead to a supernova explosion. The core collapse is clearly non-uniform. The central region collapses nearly homologously and falls away from the rest of the core. It overshoots its equilibrium point and bounces. In much less than a hundredth of a second the central region rebounds and drives outward. The piston like action of the central region of the core creates a reflected shock wave which propagates into the outer regions of the core. In the case shown in figure 1 no matter is ejected from the core. The shock wave propagates only part of the way through the core and dies out. The overlying material then rains down upon the hot neutron star at the center. No supernova results in this case.

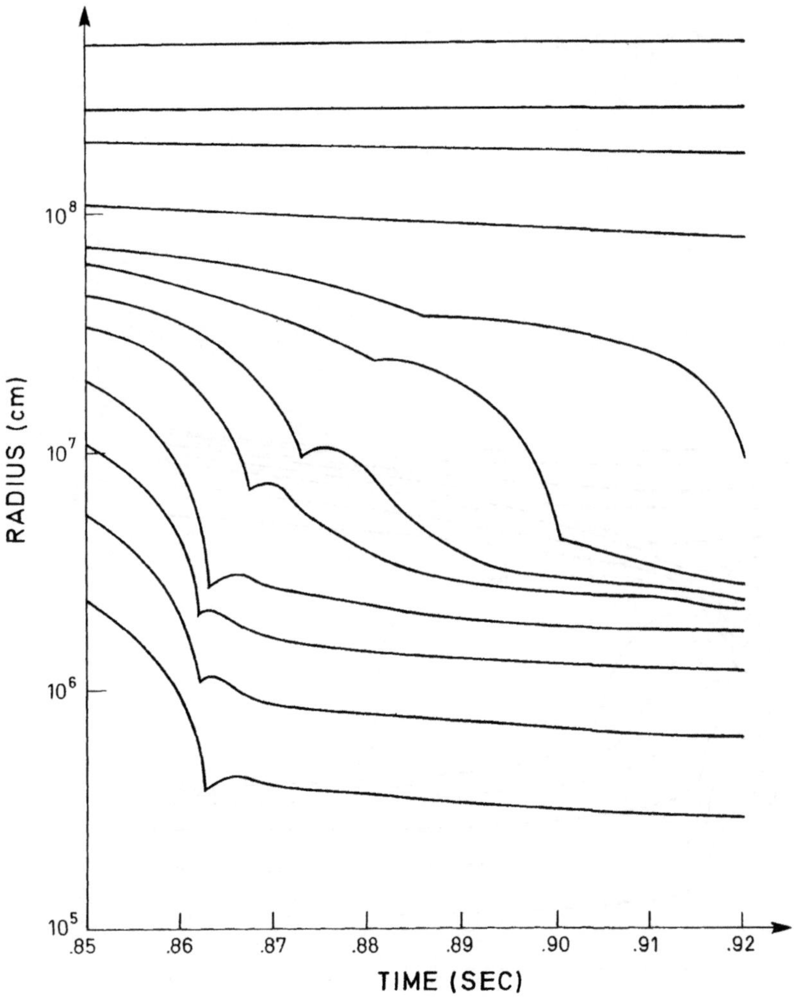

Figure 1. The evolution of a stellar core during gravitational collapse. Each line corresponds to the trajectory of an element of matter.

Figure 2 shows a model which yields an explosion. The same sequence occurs: homologous collapse of the central region of the core, rebound of this central region and the formation of a reflected shock wave. In this case, however, the shock wave maintains its strength and propagates through the entire core. At the termination of the calculations matter was moving outward with greater than the escape velocity. This model is capable of ejecting some core material and imparting energy to the overlying stellar envelope.

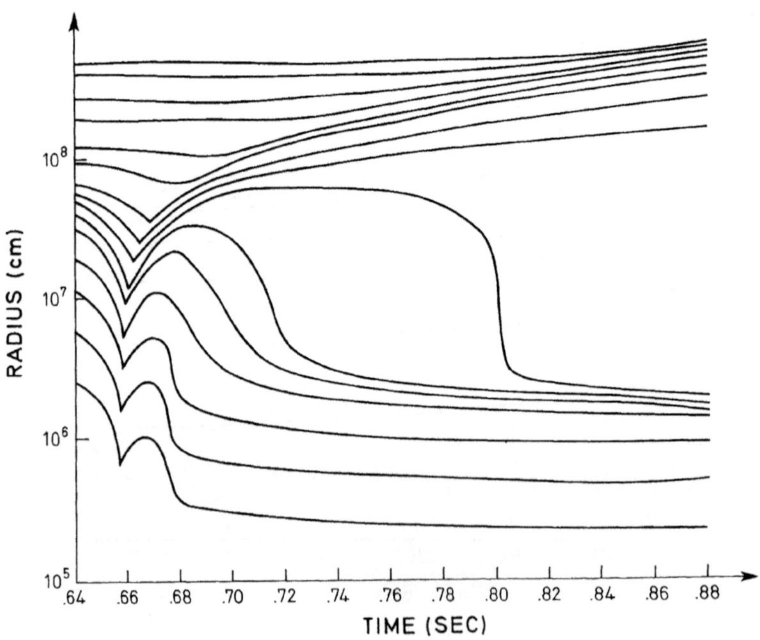

Figure 2. The same as Figure 1, except different neutrino cross-sections etc. were used.

What distinguishes between the models of figure 1 and 2? In order to discuss this question it is useful to separately consider the rebound of the central region of the core and the propagation of the reflected shock wave, and to examine the effects which neutrino production and transport has on these phenomena.

As discussed above, the core collapses due to the electron capture reactions which decrease the electron degeneracy pressure and due to the endothermic dissociation of the heavy elements. Prior to collapse the core was only marginally stable; it was mainly supported by relativistic electron degeneracy so that the average adiabatic index was just over 4/3. The decrease of the pressure in the core due to the electron captures and the dissociation of the heavy elements effectively lowers the average adiabatic index to below 4/3, and the core becomes very unstable to collapse (see Barkat, 1975). Regardless of whether any neutrinos escape or there is any other loss of energy from the stellar core, the core will contract to a more dense neutron star configuration. Furthermore the collapse is sufficiently rapid so that the core overshoots the equilibrium configuration and rebounds. Under what conditions can one expect that the rebound is especially strong? First consider an adiabatic collapse when there is no escape of neutrinos. In this case the speed of the collapse depends on the rates at which the matter is neutronized and dissociated. The inclusion of energy loss due to neutrino escape can only increase the rate of the collapse, however neutrino escape will not necessarily insure that the rebound is more vigorous. If the neutrino escape is prompt, that is if the neutrinos escape as soon as they are formed, then the collapse is as rapid as possible and the core collapse to the maximum possibly density before rebounding. This situation leads to a rapid supersonic collapse and a very strong rebound and hence a strong reflected shock wave. On the other hand if the neutrinos are partially trapped within the core so that the characteristic escape time for the neutrinos is of the same order as the time which is required for the central part of the core to collapse then a weak rebound, weaker than in the adiabatic case, results. When the neutrinos are partially trapped, they exert a great deal of pressure during the initial part of the collapse, but their contribution steadily decreases and during the reexpansion they contribute less pressure. That is, the neutrino pressure cushions the implosion more than it drives the subsequent re-explosion.

One can start to see why it is difficult to predict how a supernova model calculation will change when the input physics is varied. If the neutrino opacity is changed so that the core is first opaque and then becomes transparent, the rebound becomes weaker and then becomes stronger. Furthermore, as is discussed below, the effects of the escaping neutrinos on the dynamics of the overlying regions of the core also depend sensitively on the exact rate of which the neutrinos escape.

Now let us consider how the energy which is generated in the central region of the core can be transmitted to the outer regions of the core where it can effect the stellar envelope. Of particular importance is the role which the transport of the neutrinos play in weakening or strengthening the reflected shock wave (Bruenn et al., 1976; Schramm and Arnett, 1975).

Neutrinos which are diffusing or streaming out of the hot dense central core region can deposit energy to the outer regions by either heating this matter or simply by transferring the momentum of their streaming motion. When neutrinos Compton scatter from electrons or are absorbed by nuclei, they tend to come into thermal equilibrium with their surroundings (Zeldovich and Raizer, 1967; Tubbs and Schramm, 1975; Yueh and Buchler, 1975, 1976). Neutrino scattering from nuclei, on the other hand is essentially elastic (Freedman, 1974; Lamb and Pethic, 1976). Because of the large masses of nuclei, of the order of GeV/c^2, relative to the neutrino energies, which are typically of the order of 20 MeV, a neutrino recoils from a slowly moving nucleon or nucleus with its energy virtually unchanges. These elastic scatterings nevertheless do impart an impulse of momentum to the nucleus of the order of the neutrino momentum.

In the early supernova models (Colgate and White, 1966; Schwartz, 1967; Arnett, 1966, 1967, 1968) the driving force of the explosion was thought to be the transport of thermal energy by the neutrinos. The neutrinos which are emitted in the hot gas behind the shock wave by thermal processes (see Barkat, 1975) diffuse outward and heat up the matter in front of the shock wave. The thermal pressure then causes the outer region of the core to expand thereby initiating a shock wave in the stellar envelope which could produce the observable features of a supernova (Falk and Arnett, 1976). It should be noted that this energy deposition mechanism provides a means whereby energy is transferred from the inner regions to the outer regions of the core, but it does not help, and in fact hinders, the propagation of the reflected shock wave which formed due to the initial rebounding of the central portions of the core. Nevertheless, some recent computations have indicated that the combined effects of pre-heating the outer core regions by neutrino energy deposition and then re-heating these regions by the passage of the reflected shock wave can in some cases produce conditions which are favorable to ejecting material from the core, and presumably producing a shock wave in the envelope of the star (Bruenn, 1975; Bruenn et al., 1976).

Recently considerable attention has been given to the possibility that supernova explosions can result from neutrinos depositing momentum in the outer regions of the core. This activity has been stimulated by the developments in the theory of weak interactions and the discovery of neutral current interactions (Weinberg, 1967, 1974; Salam, 1969; Hasert et al., 1972; Benvenuti et al., 1974; Barish et al., 1974; Aubert et al., 1974). Neutral current interactions permit neutrino-nuclei elastic scattering, and if the wavelengths of the neutrinos are large compared with the dimensions of the nuclei, as they typically are in supernova models, the neutrinos scatter coherently from the nuclei so that the neutrino scattering cross section varies as the square of the mass of the nucleus. These advances in the understanding of weak interactions have led people to examine supernova models of the following type: The collapse proceeds as described above creating a dense inner core region and generating a reflected shock wave which propagates into the outer region. The inner core region which is nearly devoid of heavy nuclei has a lower opacity than the outer region which contains iron peak elements which coherently scatter neutrinos. The neutrinos are assumed to stream out of the central core where they are produced by beta and thermal processes and they are then trapped or impeded in the outer regions. This process is thought to occur simultaneously with the propagation of the reflected shock wave. The passage of the shock wave through the iron rich zones imparts an initial outward velocity, and the additional thrust of the neutrino flux further accelerates the matter to velocities which are sufficient for escape from the core. Numerical model calculations have verified that the momentum deposition mechanism can be effective if the neutrino scattering cross sections are appropriately adjusted (Bruenn et al., 1976; Wilson, private communication).

The basic phenomenology of neutrino transport supernova models is fairly well understood: the central part of the stellar core collapses and rebounds, the energy which is generated in this region is transmitted to the outer parts of the core and then to the stellar envelope either by the propagation of a reflected shock wave (which may be augmented by neutrino momentum deposition) or by the outward diffusion of energy by hot neutrinos. Why is it then that now, more than 10 years after Colgate and White suggested the neutrino transport supernova mechanism, it is still not possible to state whether this mechanism or some variation of it is capable of producing a supernova-like explosion in a realistic stellar model? This state of affairs can be ascribed to several substantial difficulties: The nature of neutrino interactions and even the number and types of neutrinos is not completely known, the properties of neutron rich matter are not adequately understood, and the numerical techniques for solving the coupled neutrino-transport hydrodynamic calculations need further development.

The cross-sections for neutrino interactions with other particles and among themselves is still uncertain. Within the Weinberg-Salam theory for weak interactions the experimental data permit about a factor of two variation in the cross-sections, and if one is not confined to this theory, then there is about an order of magnitude uncertainty. In addition Mikaelian (1976) has pointed out that besides the well known left-handed electron neutrinos and the muon neutrinos other types of neutrinos may exist and they may be important in producing supernova events. Neutrinos may be associated with the new leptons and right-handed electron neutrinos may also occur in nature. Both of these types of neutrinos could be produced in collapsing stellar cores and their production would have the effect of increasing the effective neutrino emission rate.

The structure of collapsing stellar cores, the rate of neutronization, the emission rate and spectra of the beta neutrinos and to some extent the rate of the collapse depends on the properties of the neutron rich nuclei at high temperatures. This is a difficult problem since it requires knowledge of the excited level structure of nuclei which are unstable under normal laboratory conditions, and thus requires considerable extrapolation from the properties of known nuclei (Hansen, 1968; Mazurek et al., 1974; Epstein and Arnett, 1975; Kodama and Takahshi, 1975).

Besides the uncertainties in the physical processes which are involved in supernova models, the computational techniques are not yet adequate for accurately simultating the complex behavior of the stellar cores. The hydrodynamical calculations have to be coupled to neutrino transport calculations which include the effects of neutrino emission, absorption and scattering under conditions where the electrons and neutrinos are influenced by Pauli exclusion (Mazurek, 1975; Imshennik and Nadyzhin, 1973). An accurate treatment of the transport problem can be obtained from a Monte Carlo simulation or by solving the Boltzmann equation (Yueh and Buchler, 1976b). Either approach is very time consuming when used in a dynamical calculation and some simpler yet accurate approximate techniques need to be developed.

The calculations of supernova models which have been performed so far have by necessity used various approximations for the neutrino interactions, the properties of the matter and the neutrino transport. These calculations have shown that the behavior of the core is often sensitive to relatively minor changes in the input physics in often unpredictable ways. These calcualtions have indicated which aspects of the supernova models have to be refined and they have delineated the interesting ranges of the physical parameters. Because this field is evolving so rapidly, none of the presently performed calculations have actually used all of the most sophisticated input physics which is currently available. It will be a while before model calculations are performed which employ all of these latest advances. The intensive effort which is being directed to the study of the neutrino transport supernova problem and the physics related to it has clarified the phenomena associated with the collapse of stellar cores and the production of supernova explosions. Hopefully this effort will soon yield some conclusions as to whether the neutrino transport mechanism or some related mechanism can in fact produce supernova explosions.

I wish to thank D.N. Schramm, J.R. Wilson and A.G.W. Cameron for the help they have given me in preparing this report. This work was supported in part by a U.S. National Science Foundation grant at Harvard University.

References

Arnett, W.D., Ap.J. 176, 681 (1972).

Aubert, B. et al. (Harvard-Pennsylvania-Wisconsin-FNAL collaboration, Phys.Rev.Letters 32, 1457 (1974).

Barish, S. et al. (Argonne-Concordia-Purdue collaboration), Phys.Rev. Letters 33, 448 (1974).

Barkat, Z., Ann.Rev.Ast.and Ap. 13, 45 (1975).

Barkat, S., Rakavy, G., Reiss, Y. and Wilson, J.R., Ap.J. 196, 633 (1975).

Bruenn, S., Ann.N.Y.Acad.Sci.262, 80 (1975).

Bruenn, S., Arnett, W.D. and Schramm, D.N., preprint (1976).

Colgate, S.A. and White, R.H., Ap.J. 143, 626 (1966).

Epstein, R.I. and Arnett, W.D., Ap.J. 201, 202 (1975).

Falk, S.W. and Arnett, W.D., Ap.J.Supp., in press (1976).

Fowler, W.A. and Hoyle, F., Ap.J.Supp. 9, 201 (1964).

Freedman, D.Z.,Phys.Rev.D 9, 1389 (1974).

Hansen, C.J.,Ap.Space Sci. 1, 499 (1968).

Hansen, C.J. and Wheeler, J.C.,Ap.Space Sci. 3, 464 (1969).

Hasert, F.J. et al. (Aachen-Brussels-CERN-Ecole Polytechnique - Milan-Orsay-London collaboration), Phys.Letters 46B, 138 (1973).

Kodama, T., Takahashi, K.,Nucl.Phys. A239, 489 (1975).

Lamb, D.Q. and Pethic, C.J.,Ap.J.Lett. 209, L77 (1976).

Le Blanc, J.M. and Wilson, J.R., Ap.J. 161, 541 (1970).

Mazurek, T.J., Ap.Space Sci. 35, 117 (1975).

Mazurek, T.J., Truran, J.W. and Cameron, A.G.W., Ap.Space Sci. 27, 261 (1974).

Meier, D., Epstein, R.I., Schramm, D.N. and Arnett, W.D., Ap.J. 204, 869 (1976).

Mikaelian, K.O., Phys.Rev.Lett. 36, 1089 (1976).

Paczynski, B.E., Acta Astron. 20, 47 (1970).

Salam, A., in Elementary Particle Theory, ed. N. Svartholm (Stockholm: Almquist and Wiksell), p. 367 (1969).

Schramm, D.N. and Arnett, W.D., Ap.J. 198, 629 (1975).

Tubbs, D.L. and Schramm, D.N., Ap.J. 201, 467 (1975).

Weinberg, S., Phys.Rev.Lett. 19, 1264 (1967).

Weinberg, S., Rev.Mod.Phys. 46, 255 (1974).

Wilson, J.R., Ap.J. 163, 209 (1971).

Wilson, J.R., Phys.Rev.Lett. 32, 849 (1974).

Wilson, J.R., Couch, R., Cochran, S., Le Blanc, J. and Barkat, Z., Ann.N.Y.Acad.Sci. 262, 54 (1975).

Yueh, W.R. and Buchler, J.R., Ap.Space Sci. 39, 429 (1976a).

Yueh, W.R. and Buchler, J.R., preprint (1976b).

Zeldovich, Ya.B. and Raizer, Yu.P., Physics of Shock Waves and High Temperature Hydrodynamic Phenomena (New York: Academic Press), (1967).

ASTROPHYSICS AND SPACE SCIENCE LIBRARY

1. C. de Jager (ed.), *The Solar Spectrum, Proceedings of the Symposium held at the University of Utrecht, 26–31 August, 1963.* 1965, XIV + 417 pp.
2. J. Ortner and H. Maseland (eds.), *Introduction to Solar Terrestrial Relations, Proceedings of the Summer School in Space Physics held in Alpbach, Austria, July 15–August 10, 1963 and Organized by the European Preparatory Commission for Space Research.* 1965, IX + 506 pp.
3. C. C. Chang and S. S. Huang (eds.), *Proceedings of the Plasma Space Science Symposium, held at the Catholic University of America, Washington, D.C., June 11–14, 1963.* 1965, IX + 377 pp.
4. Zdeněk Kopal, *An Introduction to the Study of the Moon.* 1966, XII + 464 pp.
5. B. M. McCormac (ed.), *Radiation Trapped in the Earth's Magnetic Field. Proceedings of the Advanced Study Institute, held at the Chr. Michelsen Institute, Bergen, Norway, August 16–September 3, 1965.* 1966, XII + 901 pp.
6. A. B. Underhill, *The Early Type Stars.* 1966, XII + 282 pp.
7. Jean Kovalevsky, *Introduction to Celestial Mechanics.* 1967, VIII + 427 pp.
8. Zdeněk Kopal and Constantine L. Goudas (eds.), *Measure of the Moon. Proceedings of the 2nd International Conference on Selenodesy and Lunar Topography, held in the University of Manchester, England, May 30–June 4, 1966.* 1967, XVIII + 479 pp.
9. J. G. Emming (ed.), *Electromagnetic Radiation in Space. Proceedings of the 3rd ESRO Summer School in Space Physics, held in Alpbach, Austria, from 19 July to 13 August, 1965.* 1968, VIII + 307 pp.
10. R. L. Carovillano, John, F. McClay, and Henry R. Radoski (eds.), *Physics of the Magnetosphere, Based upon the Proceedings of the Conference held at Boston College, June 19–28, 1967.* 1968, X + 686 pp.
11. Syun-Ichi Akasofu, *Polar and Magnetospheric Substorms.* 1968, XVIII + 280 pp.
12. Peter M. Millman (ed.), *Meteorite Research. Proceedings of a Symposium on Meteorite Research, held in Vienna, Austria, 7–13 August, 1968.* 1969, XV + 941 pp.
13. Margherita Hack (ed.), *Mass Loss from Stars. Proceedings of the 2nd Trieste Colloquium on Astrophysics, 12–17 September, 1968.* 1969, XII + 345 pp.
14. N. D'Angelo (ed.), *Low-Frequency Waves and Irregularities in the Ionosphere. Proceedings of the 2nd ESRIN-ESLAB Symposium, held in Frascati, Italy, 23–27 September, 1968.* 1969, VII + 218 pp.
15. G. A. Partel (ed.), *Space Engineering. Proceedings of the 2nd International Conference on Space Engineering, held at the Fondazione Giorgio Cini, Isola di San Giorgio, Venice, Italy, May 7–10, 1969.* 1970, XI + 728 pp.
16. S. Fred Singer (ed.), *Manned Laboratories in Space. Second International Orbital Laboratory Symposium.* 1969, XIII + 133 pp.
17. B. M. McCormac (ed.), *Particles and Fields in the Magnetosphere. Symposium Organized by the Summer Advanced Study Institute, held at the University of California, Santa Barbara, Calif., August 4–15, 1969.* 1970, XI + 450 pp.
18. Jean-Claude Pecker, *Experimental Astronomy.* 1970, X + 105 pp.
19. V. Manno and D. E. Page (eds.), *Intercorrelated Satellite Observations related to Solar Events. Proceedings of the 3rd ESLAB/ESRIN Symposium held in Noordwijk, The Netherlands, September 16–19, 1969.* 1970, XVI + 627 pp.
20. L. Mansinha, D. E. Smylie, and A. E. Beck, *Earthquake Displacement Fields and the Rotation of the Earth, A NATO Advanced Study Institute Conference Organized by the Department of Geophysics, University of Western Ontario, London, Canada, June 22–28, 1969.* 1970, XI + 308 pp.
21. Jean-Claude Pecker, *Space Observatories.* 1970, XI + 120 pp.
22. L. N. Mavridis (ed.), *Structure and Evolution of the Galaxy. Proceedings of the NATO Advanced Study Institute, held in Athens, September 8–19, 1969.* 1971, VII + 312 pp.
23. A. Muller (ed.), *The Magellanic Clouds. A European Southern Observatory Presentation: Principal Prospects, Current Observational and Theoretical Approaches, and Prospects for Future Research, Based on the Symposium on the Magellanic Clouds, held in Santiago de Chile, March 1969, on the Occasion of the Dedication of the European Southern Observatory.* 1971, XII + 189 pp.

24. B. M. McCormac (ed.), *The Radiating Atmosphere. Proceedings of a Symposium Organized by the Summer Advanced Study Institute, held at Queen's University, Kingston, Ontario, August 3–14, 1970.* 1971, XI + 455 pp.
25. G. Fiocco (ed.), *Mesospheric Models and Related Experiments. Proceedings of the 4th ESRIN-ESLAB Symposium, held at Frascati, Italy, July 6–10, 1970.* 1971, VIII + 298 pp.
26. I. Atanasijević, *Selected Exercises in Galactic Astronomy.* 1971, XII + 144 pp.
27. C. J. Macris (ed.), *Physics of the Solar Corona. Proceedings of the NATO Advanced Study Institute on Physics of the Solar Corona, held at Cavouri-Vouliagmeni, Athens, Greece, 6–17 September 1970.* 1971, XII + 345 pp.
28. F. Delobeau, *The Environment of the Earth.* 1971, IX + 113 pp.
29. E. R. Dyer (general ed.), *Solar-Terrestrial Physics/1970. Proceedings of the International Symposium on Solar-Terrestrial Physics, held in Leningrad, U.S.S.R., 12–19 May 1970.* 1972, VIII + 938 pp.
30. V. Manno and J. Ring (eds.), *Infrared Detection Techniques for Space Research. Proceedings of the 5th ESLAB-ESRIN Symposium, held in Noordwijk, The Netherlands, June 8–11, 1971.* 1972, XII + 344 pp.
31. M. Lecar (ed.), *Gravitational N-Body Problem. Proceedings of IAU Colloquium No. 10, held in Cambridge, England, August 12–15, 1970.* 1972, XI + 441 pp.
32. B. M. McCormac (ed.), *Earth's Magnetospheric Processes. Proceedings of a Symposium Organized by the Summer Advanced Study Institute and Ninth ESRO Summer School, held in Cortina, Italy, August 30–September 10, 1971.* 1972, VIII + 417 pp.
33. Antonin Rükl, *Maps of Lunar Hemispheres.* 1972, V + 24 pp.
34. V. Kourganoff, *Introduction to the Physics of Stellar Interiors.* 1973, XI + 115 pp.
35. B. M. McCormac (ed.), *Physics and Chemistry of Upper Atmospheres. Proceedings of a Symposium Organized by the Summer Advanced Study Institute, held at the University of Orléans, France, July 31–August 11, 1972.* 1973, VIII + 389 pp.
36. J. D. Fernie (ed.), *Variable Stars in Globular Clusters and in Related Systems. Proceedings of the IAU Colloquium No. 21, held at the University of Toronto, Toronto, Canada, August 29–31, 1972.* 1973, IX + 234 pp.
37. R. J. L. Grard (ed.), *Photon and Particle Interaction with Surfaces in Space. Proceedings of the 6th ESLAB Symposium, held at Noordwijk, The Netherlands, 26–29 September, 1972.* 1973, XV + 577 pp.
38. Werner Israel (ed.), *Relativity, Astrophysics and Cosmology. Proceedings of the Summer School, held 14–26 August, 1972, at the BANFF Centre, BANFF, Alberta, Canada.* 1973, IX + 323 pp.
39. B. D. Tapley and V. Szebehely (eds.), *Recent Advances in Dynamical Astronomy. Proceedings of the NATO Advanced Study Institute in Dynamical Astronomy, held in Cortina d'Ampezzo, Italy, August 9–12, 1972.* 1973, XIII + 468 pp.
40. A. G. W. Cameron (ed.), *Cosmochemistry. Proceedings of the Symposium on Cosmochemistry, held at the Smithsonian Astrophysical Observatory, Cambridge, Mass., August 14–16, 1972.* 1973, X + 173 pp.
41. M. Golay, *Introduction to Astronomical Photometry.* 1974, IX + 364 pp.
42. D. E. Page (ed.), *Correlated Interplanetary and Magnetospheric Observations. Proceedings of the 7th ESLAB Symposium, held at Saulgau, W. Germany, 22–25 May, 1973.* 1974, XIV + 662 pp.
43. Riccardo Giacconi and Herbert Gursky (eds.), *X-Ray Astronomy.* 1974, X + 450 pp.
44. B. M. McCormac (ed.), *Magnetospheric Physics. Proceedings of the Advanced Summer Institute, held in Sheffield, U.K., August 1973.* 1974, VII + 399 pp.
45. C. B. Cosmovici (ed.), *Supernovae and Supernova Remnants. Proceedings of the International Conference on Supernovae, held in Lecce, Italy, May 7–11, 1973.* 1974, XVII + 387 pp.
46. A. P. Mitra, *Ionospheric Effects of Solar Flares.* 1974, XI + 294 pp.
47. S.-I. Akasofu, *Physics of Magnetospheric Substorms.* 1977, XVIII + 599 pp.
48. H. Gursky and R. Ruffini (eds.), *Neutron Stars, Black Holes and Binary X-Ray Sources.* 1975, XII + 441 pp.
49. Z. Švestka and P. Simon (eds.), *Catalog of Solar Particle Events 1955–1969. Prepared under the Auspices of Working Group 2 of the Inter-Union Commission on Solar-Terrestrial Physics.* 1975, IX + 428 pp.
50. Zdeněk Kopal and Robert W. Carder, *Mapping of the Moon.* 1974, VIII + 237 pp.
51. B. M. McCormac (ed.), *Atmospheres of Earth and the Planets. Proceedings of the Summer Advanced Study Institute, held at the University of Liège, Belgium, July 29–August 8, 1974.* 1975, VII + 454 pp.
52. V. Formisano (ed.), *The Magnetospheres of the Earth and Jupiter. Proceedings of the Neil Brice Memorial Symposium, held in Frascati, May 28–June 1, 1974.* 1975, XI + 485 pp.

53. R. Grant Athay, *The Solar Chromosphere and Corona: Quiet Sun.* 1976, XI + 504 pp.
54. C. de Jager and H. Nieuwenhuijzen (eds.), *Image Processing Techniques in Astronomy. Proceedings of a Conference, held in Utrecht on March 25–27, 1975*, XI + 418 pp.
55. N. C. Wickramasinghe and D. J. Morgan (eds.), *Solid State Astrophysics. Proceedings of a Symposium, held at the University College, Cardiff, Wales, 9–12 July 1974.* 1976, XII + 314 pp.
56. John Meaburn, *Detection and Spectrometry of Faint Light.* 1976, IX + 270 pp.
57. K. Knott and B. Battrick (eds.), *The Scientific Satellite Programme during the International Magnetospheric Study. Proceedings of the 10th ESLAB Symposium, held at Vienna, Austria, 10–13 June 1975.* 1976, XV + 464 pp.
58. B. M. McCormac (ed.), *Magnetospheric Particles and Fields. Proceedings of the Summer Advanced Study School, held in Graz, Austria, August 4–15, 1975.* 1976, VII + 331 pp.
59. B. S. P. Shen and M. Merker (eds.), *Spallation Nuclear Reactions and Their Applications.* 1976, VIII + 235 pp.
60. Walter S. Fitch (ed.), *Multiple Periodic Variable Stars. Proceedings of the International Astronomical Union Colloquium No. 29, Held at Budapest, Hungary, 1–5 September 1975.* 1976, XIV + 348 pp.
61. J. J. Burger, A. Pedersen, and B. Battrick (eds.), *Atmospheric Physics from Spacelab. Proceedings of the 11th ESLAB Symposium, Organized by the Space Science Department of the European Space Agency, held at Frascati, Italy, 11–14 May 1976.* 1976, XX + 409 pp.
62. J. Derral Mulholland (ed.), *Scientific Applications of Lunar Laser Ranging. Proceedings of a Symposium held in Austin, Tex., U.S.A., 8–10 June, 1976.* 1977, XVII + 302 pp.

GPSR Compliance

The European Union's (EU) General Product Safety Regulation (GPSR) is a set of rules that requires consumer products to be safe and our obligations to ensure this.

If you have any concerns about our products, you can contact us on ProductSafety@springernature.com

In case Publisher is established outside the EU, the EU authorized representative is:

Springer Nature Customer Service Center GmbH
Europaplatz 3
69115 Heidelberg, Germany

Batch number: 10593059

Printed by Printforce, the Netherlands